FLOWERS: A GARDEN NOTE BOOK

Copyright © 2017 Read Books Ltd.
This book is copyright and may not be
reproduced or copied in any way without
the express permission of the publisher in writing

British Library Cataloguing-in-Publication Data
A catalogue record for this book is available from the
British Library

A Short History of Gardening

Gardening is the practice of growing and cultivating plants as part of horticulture more broadly. In most domestic gardens, there are two main sets of plants; 'ornamental plants', grown for their flowers, foliage or overall appearance – and 'useful plants' such as root vegetables, leaf vegetables, fruits and herbs, grown for consumption or other uses. For many people, gardening is an incredibly relaxing and rewarding pastime, ranging from caring for large fruit orchards to residential yards including lawns, foundation plantings or flora in simple containers. Gardening is separated from farming or forestry more broadly in that it tends to be much more labour-intensive; involving *active participation* in the growing of plants.

Home-gardening has an incredibly long history, rooted in the 'forest gardening' practices of prehistoric times. In the gradual process of families improving their immediate environment, useful tree and vine species were identified, protected and improved whilst undesirable species were eliminated. Eventually foreign species were also selected and incorporated into the 'gardens.' It was only after the emergence of the first civilisations that wealthy individuals began to create gardens for aesthetic purposes. Egyptian tomb paintings from around 1500 BC provide some of the earliest physical evidence of ornamental horticulture and landscape design; depicting lotus ponds surrounded by symmetrical rows of acacias and palms. A notable example of an ancient ornamental garden was the 'Hanging Gardens of Babylon' – one of the Seven Wonders of the Ancient World.

Ancient Rome had dozens of great gardens, and Roman estates tended to be laid out with hedges and vines and contained a wide variety of flowers – acanthus, cornflowers, crocus, cyclamen, hyacinth, iris, ivy, lavender, lilies, myrtle, narcissus, poppy, rosemary and violets as well as statues and sculptures. Flower beds were also popular in the courtyards of rich Romans. The Middle Ages represented a period of decline for gardens with aesthetic purposes however. After the fall of Rome gardening was done with the purpose of growing medicinal herbs and/or decorating church altars. It was mostly monasteries that carried on the tradition of garden design and horticultural techniques during the medieval period in Europe. By the late thirteenth century, rich Europeans began to grow gardens for leisure as well as for medicinal herbs and vegetables. They generally surrounded them with walls – hence, the 'walled garden.'

These gardens advanced by the sixteenth and seventeenth centuries into symmetrical, proportioned and balanced designs with a more classical appearance. Gardens in the renaissance were adorned with sculptures (in a nod to Roman heritage), topiary and fountains. These fountains often contained 'water jokes' – hidden cascades which suddenly soaked visitors. The most famous fountains of this kind were found in the Villa d'Este (1550-1572) at Tivoli near Rome. By the late seventeenth century, European gardeners had started planting new flowers such as tulips, marigolds and sunflowers.

These highly complex designs, largely created by the aristocracy slowly gave way to the individual gardener however – and this is where this book comes in! Cottage Gardens first emerged during the Elizabethan times, originally created by poorer workers to provide themselves with food and herbs, with flowers planted amongst them for decoration. Farm workers were generally provided with cottages set in a small garden—about an acre—where they could grow food, keep pigs, chickens and often bees; the latter necessitating the planting of decorative pollen flora. By Elizabethan times there was more prosperity, and thus more room to grow flowers. Most of the early cottage garden flowers would have had practical uses though —violets were spread on the floor (for their pleasant scent and keeping out vermin); **calendulas** and **primroses** were both attractive and used in cooking. Others, such as **sweet william** and **hollyhocks** were grown entirely for their beauty.

Here lies the roots of today's home-gardener; further influenced by the 'new style' in eighteenth century England which replaced the more formal, symmetrical 'Garden à la française'. Such gardens, close to works of art, were often inspired by paintings in the classical style of landscapes by Claude Lorraine and Nicolas Poussin. The work of Lancelot 'Capability' Brown, described as 'England's greatest gardener' was particularly influential. We hope that the reader is inspired by this book, and the long and varied history of gardening itself, to experiment with some home-gardening of their own. Enjoy.

PLATE I

⅔ natural size   EUCRYPHIA PINNATIFOLIA   28 August, 1920

FLOWERS
A GARDEN NOTE BOOK

WITH SUGGESTIONS FOR GROWING THE
CHOICEST KINDS

BY THE
R<small>IGHT</small> H<small>ON</small>. SIR HERBERT MAXWELL
B<small>T</small>., F.R.S., LL.D. (G<small>LASGOW</small>), D.C.L. (D<small>URHAM</small>), V.M.H.

*WITH TWELVE COLOURED PLATES FROM DRAWINGS
BY THE AUTHOR*

"Tantus amor florum!"

THESE PAGES ARE INSCRIBED TO

ARTHUR GROVE

IN REMEMBRANCE OF MANY HOURS
PROFITABLY PASSED WITH HIM IN
SUNSHINE AND SHOWER

Contents

		PAGE
To the Reader		xi
I. Some Hardy Bulbs		1
II. The Herbaceous Border		27
III. Some Flowering Shrubs		53
IV. Some Rhododendrons		102
V. Wild Gardening		127
VI. The Choice of Plants		147
VII. Some Plants for Walls		166
VIII. Rockwork and Edgings		181
IX. Some Failures		193
X. Some Weeds		208
XI. Some Plant Names		216
XII. L'Envoi		233
General Index		237
Index of Plant Names		241

Illustrations in Colour

		PAGE
I.	Eucryphia Pinnatifolia	*Frontispiece*
II.	Lilium Regale	20
III.	Lilium Browni	24
IV.	'Lady's Slipper' (Cypripedium Calceolus)	40
V.	Meconopsis Simplicifolia (Bailey's Variety)	52
VI.	'Dragon's Mouth' (Helicodiceros Crinitus)	56
VII.	Desfontainea Spinosa	80
VIII.	'Kowhai' (Clianthus Puniceus)	88
IX.	Buddleia Colvillei	92
X.	Rhododendron Soulei	108
XI.	Alstroemeria Hookeri	164
XII.	Pæonia Cambessodessi	208

To the Reader

IT used to be commonly said that an Englishman's first words of a morning were " What shall we kill to-day ? " It would ill become one who, like the writer of the following pages, has in his time put an end to the lives of a considerable number of his furred, feathered and scaly fellow-creatures, to join in denouncing what the Humanitarian League term " blood sports " ; but whereas during the later part of a long life his more frequent morning thought has been—" What shall we *plant* to-day ? " it is possible that among the notes that have accumulated in the lapse of seasons there may be something that may serve, not to instruct, but to encourage the enterprise of other amateurs.

When my departed friend Andrew Lang described gardening as " a device of Providence for the pottering peace of virtuous eld," I, being then some thirty years younger than I must own to now, rebuked him for not writing with more respect for the most ancient of all industries [1]—the most enduring of all pastimes.

[1] Probably it is not the most ancient. Our earliest ancestors seem to have supported themselves and their families by the chase long before they took to tilling the ground.

To the Reader

Howbeit, I have now lived long enough to confirm the efficacy of Lang's prescription, without endorsing his limitation thereof to "virtuous eld." He, at all events, was a kindlier critic than the cynic who penned the elegiac couplet—

> To every man comes a time when his favourite sins all forsake him,
> And he complacently thinks he has forsaken his sins.

It is told of the novelist, Alphonse Karr, that when he ceased writing fiction and took to discoursing about his garden, somebody asked him the reason why. "When I was young," he replied, "I lived romance and I wrote romance. I am now a gardener and I write about gardens." I am not conscious of having "lived romance"; perhaps had I done so the few novels I wrote *dans le temps* might have proved more popular than they did, and as for gardening literature, its output has grown so prodigiously of late years that one may well hesitate before adding to it. I have so hesitated; but whereas in the matter of "talking shop," the votaries of gardening have their equals only among those of golf and money-making, I have yielded to a temptation to record a chequered experience of failure and success with some of the less common shrubs and herbs that may be grown in our fickle British climate. The interest of out-of-door gardening has been so greatly enhanced within the last thirty years by the vast number of new species introduced to this country from the far east by such intrepid collectors as Wilson, Forrest, Henry, Farrer, Kingdon

To the Reader

Ward, Purdom and others—the abundance at an amateur's disposal is now so varied—that the difficulty consists, not in acquiring material to furnish borders and adorn woodland withal, but in discriminating between what is good and not so good—in choosing the best species and rejecting all of inferior merit. My readers, therefore, will please not to expect to find in these desultory chapters a treatise on garden craft in general or on the mysteries of propagating, manuring, trenching, pruning, etc., in particular; they contain nothing of more importance than the incidental kind of talk that occurs in rambling round a garden with a friend interested in flowering things. Their aim is only serious in so far as it is an attempt to indicate certain points in the arrangement of shrubs and the treatment of flower-borders, which are not so commonly kept in view as they might be with advantage, and to furnish a general reply to some, at least, of the enquiries which the writer has had to answer so frequently in detail—what are the best things to plant? Both in the text and in the coloured plates I have endeavoured to draw attention to plants out of the ordinary run of nursery stock. Nurserymen, of course, have to be prepared to meet whatever may be the prevailing demand, and nobody has a right to complain if he cannot be supplied at once with some plant out of the common. The wonder is that so many nurserymen grudge neither time nor expense in growing plants for which the demand must be intermittent and uncertain. Luckily for amateur gardeners, there

To the Reader

are as keen enthusiasts in the trade as there are outside it.

I have endeavoured to indicate the relative hardiness of the plants mentioned, but whereas my experience of them has chiefly been gained in a mild, maritime district, I trust it may be taken subject to that limitation. There are, of course, many plants which, in inland regions, will not pass through a hard winter in the open, but are quite capable of doing so when trained on a wall or planted at its foot. The chief troubles that beset the cultivator in all parts of Great Britain and Ireland is not severe winter cold, but spring frosts and violent winds.

It will be understood, I hope, that no attempt has been made to give in the following pages anything approaching a full list of desirable plants. That could only be the outcome of experience in a botanical garden. My purpose has been no more ambitious than to offer suggestion for choice from the abundance of good things within easy reach of any amateur. Some surprise may be felt at the absence of any reference to garden roses, carnations, and other florists' flowers; the omission is not owing to want of esteem for such things; quite the contrary, for although I confess to prepossession or prejudice in favour of natural species of plants rather than those artificially bred by man, I cannot go all lengths with Perdita in denouncing hybrids :

> ... The fairest flowers o' the season
> Are our carnations and streak'd gillyflowers,

To the Reader

> Which some call nature's bastards. Of that kind
> Our rustic garden's barren ; and I care not
> To get slips of them. . . .
>
> . . . . I'll not put
> The dibble in earth to set one slip of them.

To bar such fragrant treasures as hybrid tea roses and clove carnations would be to despoil the flower garden of its chief attraction in summer and turn it into a mere playground for pedantic botanists like myself. Such is far from the intention in these pages ; but it is prudent to leave the production and cultivation of florists' flowers to be dealt with by more competent counsellors than the present writer. It is surely well that books on gardening, like gardens themselves, should not be designed on a uniform plan ; and it may serve to screen some of the shortcomings in the following essay if I quote thus from wise old Robert Burton's *Anatomy of Melancholy* :

"The very being in the country, that life itself, is a sufficient recreation to some men, to enjoy such pleasures, as those old patriarks did. Dioclesian the emperor was so much affected with it that he gave over his scepter and turn'd gardener . . . If my testimony were ought worth, I could say as much of myself ; I am *vere Saturninus* ; no man ever took more delight in springs, woods, groves, gardens, walks, fishponds, rivers, etc."

Except Plates X and XI, the illustrations have been necessarily reduced in scale from water-colour drawings the size of nature of some flowers grown in the open at Monreith.

I

Some Hardy Bulbs

> ... O Proserpina,
> For the flowers that, frighted, thou let'st fall
> From Dis's waggon ! Daffodils
> That come before the swallow dares, and take
> The winds of March with beauty ...
> The crown-imperial ; lilies of all kinds,
> The flower-de-luce being one ! O these I lack
> To make you garlands of ... Come, take your flowers.
> *Winter's Tale*, iv. 3.

AMONG plants popularly known as bulbous are included those which grow from corms, tubers and rhizomes, as well as those which have true bulbs. A true bulb, such as that of a snowdrop, tulip or lily is really a subterranean bud consisting of fleshy leaves closely packed round a woody core, whence roots proceed downward and the flowering stem upward. A bulb is perennial, living for an indefinite number of years.[1] A corm, such as that of a crocus, a gladiolus or a montbretia, differs from a bulb in being

---

[1] The bulb of *Lilium giganteum* may seem to be an exception to this rule, forasmuch as it dies after flowering ; but it is no exception, for it lives an indefinite number of years before throwing up a flowering stem, leaving, when it dies, a number of young bulbs to take up the running.

Flowers

solid, with no trace of imbricated leaves or scales, and in being annual, not perennial. The corm of a crocus planted in autumn will die after flowering in spring, a new corm forming on the top of it, which in turn will die after flowering in the following spring, and so on *ad infinitum*. Such a process must inevitably bring the new corms to the surface of the ground, were it not for a peculiar and very interesting contrivance to keep them buried. Each new corm sends out two kinds of root—namely, the ordinary fibrous roots which absorb nourishment for the plant, and roots of another character and special function. The latter kind, termed "contractile" roots, push their way into the soil deeper than the old corm, anchor themselves there, and then contract, dragging the new corm down to a safe depth, sometimes to a lower level than the old one. They may be distinguished from the fibrous roots by their greater thickness and by the ringed and wrinkled appearance.

As for tubers and rhizomes, it may only be noted here that tubers are merely a swelling of the stem, sometimes furnished with "eyes" or buds, as in the potato, the anemone and the winter aconite; at other times without "eyes," as in the turnip, the cyclamen and the herbaceous peony. A rhizome is another form of thickened stem, generally spreading horizontally along or under the surface of the soil, as in the lily of the valley and the German iris.

Apology is due, and is hereby respectfully offered, to those of my readers to whom all this is as familiarly known as it is to the writer; but many good amateurs

Some Hardy Bulbs

make considerable progress in gardening before giving attention to the biology of plants. In what follows, bulbs and corms shall be treated together; tubers and rhizomes being left out of account for the present.

There is no bulbous plant known to me which, being of extreme hardihood in resisting cold, shows such marked preference or distaste for certain parts of our country as does the common snowdrop. A distance of but a few miles sometimes makes much difference in its behaviour, even though there may be little variation in soil or climate. Take, for instance, two places on the Firth of Clyde in the same county of Renfrew. In Sir Hugh Shaw Stewart's romantic demesne of Ardgowan the woods are sheeted with snowdrops in measure that may be estimated by the acre; while in Sir John Stirling Maxwell's grounds at Pollok, distant only some four-and-twenty miles, it is difficult to keep them alive and impossible to make them increase. I am speaking only of the common snowdrop, *Galanthus nivalis*, which, in my humble judgment, will brook comparison with any other species or variety when growing in congenial environment.

It is well known what splendid consignments of early bulbous flowers are sent to London and other great towns from that paradise of horticulture, the Scilly Isles. It happened some years ago that these consignments began to arrive much in advance of the usual date. Having noted that the common snowdrop is not affected by temperature in the same way that other plants are in respect of their date of flowering,

Flowers

I asked the late Mr. Dorrien Smith—the originator and organiser of the Scilly Islands flower industry—whether his snowdrops were flowering earlier than usual in that season. "Snowdrops!" he replied, "they won't grow in Scilly. They don't like us."

This sounded strange to one living, as I do, in a district where snowdrops carpet the woods as lavishly as the common blue hyacinth.

Now, in putting that question to "the King of Scilly," I was assuming that snowdrops flourished in his realm as freely as they do on our more northerly west coast, and I had in mind the peculiar temperament of this little plant which causes it to refuse to be forced into bloom. Narcissus, crocus, squills and tulips respond readily to artificial heat, but the snowdrop—NO! It is true that if the ground outside is frost-bound, rendering it physically impossible for the leaves that enclose the flower-buds to push through, snowdrops grown under glass and free from the impediment of frozen soil will get a start over their sisters in the open, but not a day before the time appointed for that particular season. Our garden book records the date of the first snowdrop to flower in the open in each of the last seventeen years, and it has not been possible to trace any connection between that date and the character of the season. For instance, the winter of 1922-3 was the mildest in my long recollection, yet no snowdrop drooped from its stalk till 5th January; whereas in 1921-22 the earliest to flower was on 22nd December. It is good to note how the points of the leaves enclosing the flower-buds

Some Hardy Bulbs

are thickened and toughened at the tips, the better to enable them to thrust through the soil. This simple armature shows on the mature leaf like a delicate nail on a green digit.

The snowflake—*Leucojum vernum*—usually comes into bloom a fortnight or three weeks later than the snowdrop; but we have had the Carpathian variety, *L. vernum Vagneri*, with two heads on the stem instead of one, in flower as early as 4th January. It is a beautiful and fragrant thing, and, being rabbit-proof, colonises a woodland as readily as the snowdrop; indeed it grows freely in some districts where snowdrops have to be coddled. Howbeit, snowdrops and snowflakes should not be planted in mixture, but kept in separate glades, for their flowering season overlaps, and they resemble each other so nearly that they are better kept apart.

From these forerunners of spring the transition to narcissus is easy; but so numerous are the natural species—so vast the multitude of hybrids and varieties created by busy-fingered florists—that I can but mention a few that I hold in special esteem. Many years ago—'twas in the early 'eighties—I served on the Narcissus Committee of the Royal Horticultural Society. Even in those distant days the annual production of fresh varieties was so profuse that I grew bewildered, wearied and resigned. Truth to tell, my heart goes out far more readily to a natural species and its spontaneous varieties than to the more showy, but often coarser, manipulated products of the florist. One must not grudge enthusiasts the

Flowers

excitement of creating new forms of a beautiful flower, still less envy them the reward of their skill, which runs sometimes, I am told, to the handsome figure of £50 for a single bulb.[1]

The primary merit of a flower in its appeal to human perception being beauty, none of the myriad forms into which the Magni-coronati or Trumpet section of *Narcissus* has been coaxed can be deemed to excel our common wild daffodil, *N. pseudo-narcissus*, and its natural varieties, *bicolor*, *minor* and *minimus*. Nevertheless, the most exclusive collection should also embrace *N. Johnstoni* " Queen of Spain," reputed to be a natural hybrid between *N. bicolor* and *triandrus*, originating in the mountains of Portugal.

In the Medio-coronati or Incomparabilis section one cannot but recognise notable enhancement of grace and refinement of colour in some of the modern forms. Confine me to a single variety in this section for planting in woodland or pasture and I should plump for *Barri conspicuus*.

In the Parvi-coronati section also the florist has wrought wonders with the fragrant polyanthus narcissus, *N. tazetta*, some charming varieties having been evolved. In mild districts they rank among the choicest border flowers and take care of themselves; but we are told that in the colder parts of Great Britain they should have the protection of a cool

---

[1] I believe the tune was set for exorbitant prices when, several years ago, Mr. Pope paid £100 for three bulbs of *Narcissus* " Will Scarlett." It was stated in the *Times* of 16th May, 1923, that the bulbs of a new variety of *N. poëticus ornatus*, occupying a plot of ground at Spalding, Lincolnshire, the size of a billiard table, had been sold for £1000.

Some Hardy Bulbs

greenhouse. In the pheasant's-eye narcissus, *N. poeticus*,[1] excellent results have followed upon breeding from the best flat-petalled forms, and care should be taken only to grow the fairest of them, whereof that known as *ornatus* may be reckoned the type. One would like to see this charming flower grown more often in the manner for which it is so admirably adapted—namely, in broad drifts in parks and pleasure grounds. It colonises as freely as the common daffodil, flowering a month or six weeks later. In the present year—1923—a sheet of the common *N. poeticus* upwards of one hundred yards long, in the park at Monreith, remained in beauty till the end of June.

Among other natural species the jonquil—*N. jonquilla*—and the campernelle—*N. odorus*—are indispensable for their fragrance; *N. cyclamineus* and *triandrus* for their jauntily reflexed corolla, and the hoop petticoat—*N. bulbocodium* for its quaint shape. Of the last-named species the sulphur-coloured variety *citrinus* takes more kindly to British conditions than the golden-flowered type, becoming naturalised under favourable conditions and spreading pretty freely by seed.

The ease with which nearly all species of narcissus can be grown in British gardens seems the more remarkable because, being mostly natives of Southern Europe and North Africa, they encounter in these

---

[1] Mr. Weathers tells us in his *Bulb Book* that this is not the flower described by Ovid (*Metamorphoses*, iii. 346), but *N. tazetta*. He does, not, however, give his authority.

Flowers

islands very different conditions of soil and climate. No genus of plants lends itself more generously to cultivation, and perhaps none is so little liable to disease. Moreover, it is protected against the assault of browsing and gnawing animals by the presence in its leaves and roots of crystals of calcium oxalate, technically termed *raphides*. These needle-shaped bodies are distributed in innumerable bundles throughout the tissues of the plant, rendering it indigestible, and possibly poisonous, to cattle and smaller mammals. Unluckily for us, the crocus, belonging to the Iris order, and the tulip, belonging to the Lily order, are not equipped for defence in this manner as are the narcissus and snowdrop, members of the Amaryllis clan.

Of the only enemy from which the narcissus seems to suffer—the fly *Merodon equestris*—I have no experience, and can only speak from hearsay. This bee-like fly, whereof the grub appears to be the only creature capable of digesting daffodils, lays an egg in or near the bulb which the said grub enters and destroys. It has wrought serious mischief from time to time in Holland and in the southern counties of England; but I have not heard complaints about it in Scottish gardens.

The well-merited popularity of the genus *Narcissus*, the readiness with which many of its species become naturalised, and the great preponderance of yellow in its livery, have given rise to an impression that yellow prevails more among spring flowers than among those of any other season. This impression

Some Hardy Bulbs

has no doubt been strengthened in our own country by the far-flung glory of the gorse, overlapping the later radiance of the broom; but I do not think it is well founded. Yellow is the commonest colour in flowers at all seasons in the northern temperate zone, a phenomenon which the late Grant Allen interpreted as the result of petals being expanded stamens; and whereas yellow is far the commonest colour of stamens, anthers and pollen, yellow flowers greatly preponderate over those of any other hue. White, the effect of the discarding of pigment, is nearly as common as yellow, while the development of pigment brings about progressive evolution of orange, red, purple, and, ultimately, blue—less frequent than any other colour among flowers.[1]

The Dutch crocuses, which are poured into this country every autumn by the million, are mainly cultivated varieties of *Crocus vernus*, and, gorgeous as is the display of which, when rightly used, they are capable, I cannot esteem them more highly than the original species. February, more often than not, is grim and grey in our northerly shire; but there are sun-lit noons even in that month, when, beneath the bare ash trees clustered round our kirkyard and manse, this common crocus completely covers the sward with a carpet of soft purple, whereof the memory shall outlive that of many showier scenes. Were it not that the crocus is the accursed rabbit's special spring

[1] The late Mr. Mangles, eponomus of a group of white-flowering rhododendrons, observed that, while plants with coloured flowers frequently sported into white, he knew no instance of a normally white flower sporting into any colour.

Flowers

delicacy, what exquisite spreads of colour might we not create in woodland glades and on water-sides!

I am not qualified to say much about other species of crocus. They are, of a truth, a fascinating family; any one who aims at intimacy with them may be well advised to study the fifth chapter of *My Garden in Spring*, by my good friend, Mr. E. A. Bowles. He has made a special study of the genus, and deals in another volume—*My Garden in Autumn*—with the autumn-flowering crocuses, a group which, in most gardens, receives far less attention than is its due. My limited experience with them enables me to endorse his advice that the easiest kinds to begin with are *C. speciosus, zonatus* and *longiflorus*, the first of this trio being the pick of the basket. Planted by the hundred in grass (and the bulbs are so cheap as to be at command of the most modest exchequer) the effect of the bluish-violet blossoms and rich orange anthers and stigmata on a sunny September afternoon is well worth some trouble to secure. And the only trouble is this—that whereas the leaves of *C. speciosus* push up in spring, the grass must not be mown till these have faded.

One often hears *Colchicum* spoken of as an autumn crocus; but the crocus is of the Iris order and has but three stamens, while colchicum is of the Lily order with six stamens. All species of colchicum are best grown in grass; their massive foliage, thrown up in spring protects them from scythe or mowing machine, but makes an untidy mess in the borders when withering in June. No animal will touch colchicum—not even the clandestine slug or the rapacious

Some Hardy Bulbs

rabbit—so it needs no protection; but it should not be planted in pasture, for it spreads rapidly both by offsets and seed, the latter being formed underground in large capsules which are thrust up with the leaves in spring. *C. speciosum* is the finest of the genus, and to see its white variety pushing stainless chalices through the dark earth is a special autumn treat.

The beauty of narcissus, crocus, squills and the like is so infinitely enhanced by planting them in grass, especially on sloping ground, that it is distressing when an effect so easily secured is missed by the bulbs being set in clumps at regular intervals or dotted as evenly as the pattern of a carpet. If they are flung broadcast and planted with a dibble where they fall, the best result will follow.

The Star-of-Bethlehem family deserves more popularity than it has received. One seldom sees in gardens any except the common *Ornithogalum umbellatum*, which is the least ornamental of the genus and is apt to become a troublesome weed. *O. nutans* and *pyramidale* are quite as easy and far prettier, and in mild districts *O. Arabicum* and *thyrsoides*, both handsome species, may be grown.

There is a queer bit of lore associated with these plants. We read in 2 Kings, vi. 25 how, during the siege of Samaria by Benhadad, King of Syria, the famine became so desperate that " an ass's head was sold for four-score pieces of silver, and the fourth part of a cab of doves' dung for five pieces of silver." Now I never could imagine what nourishment a starving population could derive from doves' dung, until the

Flowers

late Canon Tristram, of Durham, who had travelled much in Asia Minor and wrote *The Land of Moab*, explained the mystery. He told me how the plains of Syria and Palestine are sheeted in spring with the white flowers of a species of Star of Bethlehem, the bulbs of which are used as food. The Greek name for this plant—ὀρνιθόγαλον, latinised *ornithogalum*—means " birds' milk," alluding to the sheets of white blossom; but the Arabs have a less poetical name for it meaning " doves' dung." It was the roots of this plant that were so highly prized during the siege. When the Revised Version of the Old Testament was published, I turned up the passage to see whether the learned men who had worked so hard to improve the Authorised Version had detected the trap. They had not. There was the " doves' dung " as before; but " cab " had been recast as " kab! "

I have often had it in mind to sample a dish of Star-of-Bethlehem roots, whereof there is abundance in our woods; but have never been able to decide what may be the proper season for them. In clearing out a border over-run with *Alstrœmeria* I was struck by the abundance of its succulent roots which are said to be prized for food by South American Indians; so I asked my old and far-travelled friend, H. J. Elwes (now, alas! no more) whether he had ever tried them. " No," said he, " and so long as I can get a decent potato I don't intend to try them! "

It is a fertile source of speculation how primitive man ascertained what plants and fruits were wholesome and what were poisonous. How did he come to

Some Hardy Bulbs

esteem onions, potatoes and lily roots (Asiatics eat regularly the bulbs of tiger and auratum lilies) and to avoid those of narcissus, colchicum and snowdrop? Who first was so bold as to partake of horse-radish sauce? and what calamity warned him against the root of aconite? I suppose that in primitive communities experiment was made on prisoners of war and poor relations.

But I must stick to my text, or this chapter will run to unconscionable length, and, even with the utmost economy of words, I cannot do more within reasonable limits than notice a mere fraction of the desirable bulbs that may be classed as hardy. The splendid series of bulbous Iris must be passed over with but a lingering glance at *I. reticulata* in the rich purple and gold raiment with which it dares the February gale; at the many varieties of Spanish iris—*I. xiphium*—whereof the delicate grace has been sacrificed for size in the new race of Dutch irises; and at the later and equally lovely so-called English iris—*I. xiphoides*, all of these being of easiest character.

Anyone who has seen the meadows at Iffley, near Oxford, in late April, or those near Azay-le-Rideau in Touraine earlier in that month, crowded with the chequered bells of the common fritillary—*Fritillaria meleagris*, may well have wondered why such a choice display has not been prepared in other districts. There is only one hindrance to doing so, for the plants, once established, increase rapidly by seed. That hindrance exists in that child of Belial—the rabbit. Where rabbits come not, plant plenty of fritillaries—

Flowers

plum-coloured, brown and white—and you shall not miss a rich reward. For borders, there are no more refined ornaments than *Fritillaria pyrenaica* and *pallida,* the former being a flower of singular grace of form. I wish I could report favourably of *F. pudica* and *aurea*—charming little people which we have never succeeded in persuading to abide with us more than a season or two.

No border plant known to me responds so suddenly and swiftly to the stir of a still-distant spring than does the Crown Imperial—*F. imperialis.* Long before we well-clad mortals are sensible of any relaxation of winter chill and gloom—nay, sometimes when the chill and gloom are at their worst—this gallant plant thrusts fat ruddy noses through the soil, which rush with amazing rapidity into shining green columns. Lilies and other bulbs awake with deliberation, but the crown imperial rises headlong. No February would I deem to have filled its part in the calendar if it did not present us with this cheering phenomenon. What a fuss the gardening papers would be making about this great fritillary if it had lately been discovered and landed here by Mr. Wilson, Mr. Forrest, or some equally intrepid collector! Yet in comparatively few gardens is a place given to this jocund herb. In one respect it deserves forbearance which it does not always receive. After the flower is past, the tall stems wither slowly and fall over in dishevelment, suggesting their removal; but they must on no account be cut over till quite dead. Lilies of all kinds suffer amputation with impunity at any stage of

Some Hardy Bulbs

growth; not so the crown imperial, for its bulb, if shorn of its stalk, will not flower in the following seasons.

We have not yet hit upon a good English name for *Chionodoxa*, the literal translation, " glory-of-the-snow " being too pompous for everyday use. Be the frost never so keen, the wind never so bitter, the rain never so ruthless, these little flowers never bow their bright heads. They are the very Mark Tapleys among bulbous plants. The variety which goes by the preposterous title of *gigantea* requires a couple of seasons before it does itself justice; then, if care has been taken to extirpate the few bulbs that produce slatey-grey or pinkish flowers, you shall be rewarded by a constellation of sky-blue stars, whether in the border or, better still, scattered through sloping turf. There are two or three other varieties of the type which I am not learned enough to distinguish with precision. I collected some with deep blue, rather small flowers in a copse near Cettinge, the erstwhile capital of the gallant little kingdom of Montenegro, now merged in the newly-forged state of Yugo-Slavia.

The squills—*Scilla*—are distinguished from *Chionodoxa* chiefly by the stamens, which are thread-like filaments, whereas those of *Chionodoxa* are flattened and gathered into an erect cone. One cannot plant too many squills, either in the grass or as border-edgings. *S. bifolia* leads the procession, sometimes flowering before the end of January; *S. Siberica* follows in March, a charming companion for *Narcissus minor* or *nanus* (I know not which), and our native

Flowers

S. verna takes up the running in April. The last-named species abounds in the short, wind-swept turf of our seaside " heughs," growing only four inches high, and not flowering before the end of May ; whereas, brought into the shelter of the garden, it flowers from the end of March and rises to a height of eight inches. The Spanish squill—*S. Hispanica* or *campanulata*—is as tall as our wild blue hyacinth and every whit as desirable, spreading freely when planted out in the woods, as it should be, in all its colours— blue, pink and white.

The common dogtooth violet *Erythronium dens-canis* would be well worth growing for the beauty of its leaves, let alone the flowers whereof there are some nice varieties ; but these are far excelled in grace and colour by the blossoms of some of the American species —*E. Johnsoni* and the kind known as Pink Beauty, geographical variants from the Californian *E. revolutum*. The genus was given the name *Erythronium*—signifying the Ruddy One—before European botanists had become acquainted with the yellow adder's tongue— *E. Americanum*.

As for tulips, about the innumerable host of florists' varieties I will say no word, for they are mostly used for transient effect—a sort of horticultural window-dressing which bravely lights up the suburbs and deserves all commendation for so doing. More power to those who provide such admirable material ! Most of the choice natural species—*T. Greigi, Tubergiana, Clusiana, præstans* and such like seem to pine for more sun than we can give them in Scotland. They

Some Hardy Bulbs

flower brilliantly for one season, or perhaps two, after which nothing is generally left but the label. *T. Kaufmanniana*, earliest of all, is more permanent, while *T. Persica* and *sylvestris* and the charming brood of Darwin tulips make themselves quite at home and increase year by year if left alone. But I have not persevered enough to enable me to speak usefully about this magnificent race, having neither labour nor patience to fuss about bulbs that require to be lifted and stored. It is probably safe to treat all tulips with lime, and I do not know any bulbous plant that does not respond gratefully to a dose of wood ashes.

We come now to a group of plants in the cultivation of which greater uncertainty has to be faced—more frequent failure to be endured—than with any other kind of bulbous herb. Yet it is one wherein success receives richer reward than with any other. I speak of the true lilies, a genus consisting of less than one hundred known species, all natives of the northern hemisphere. It is a mournful fact that, of the immense consignments of lily bulbs of the finer Asiatic species that reach this country every year, and the less numerous species imported from North America, only a trifling percentage come to any good with us. Some may flower once from the nutriment stored within the bulb, but few survive to a second season. If only five per cent. of the lily bulbs imported annually were to flourish and increase in this country, British gardens by this time would present a display in summer transcending the dreams of the most sanguine amateur. That a far larger proportion than this might and ought

Flowers

to survive the trying ordeal of transport is certain, provided (1) that better care be applied to raising and packing the bulbs for export, and (2) that their wants on arrival should be better understood and provided for.

Take as a conspicuous instance *L. auratum*, the golden-rayed lily of Japan. That it is capable of flourishing permanently and increasing under British conditions is manifest from what may be seen in the garden of a neighbour of my own—Major Hamilton of Craighlaw. About twenty years ago—rather more than less—a chance lot of bulbs were planted in a row along a border about fifty yards long, without any precaution against disease or special preparation of soil. They have thriven vigorously, forming substantial clumps averaging five feet high. Why is this an unusual experience? Why is the result so different in my own garden, only a dozen miles from the other? We have lost nearly every one of the scores of *auratum* whereon I have wasted my substance, while those we have raised from scales are all that could be desired. By that means—breaking up a sound bulb and setting the scales, a good and trustworthy stock can be raised in any quantity required, and the bulbs so produced have proved, without exception, to be as hardy and permanent as those of tiger lilies. So, probably, are bulbs raised from seed, when that can be had, but the process is much slower.

The primary cause of failure with imported bulbs must be assigned to the system pursued by Japanese growers, who hurry the bulbs into exorbitant growth

Some Hardy Bulbs

with stimulating manure without rotation of crop, sometimes shearing off the basal roots before packing for export. It is on these roots that the future nourishment of the bulb depends. When re-planted in this country, it may send up a strong flowering stem nourished by the annual roots produced above the bulb; but the bulb itself generally dies after that effort, having no basal roots to maintain and renew its vigour. Nor is that the whole of the evil result of over-propagation, forcing cultivation and faulty packing. The great majority of the bulbs are found on arrival to be infested either with mites or with the fungus *Rhizopus necans*, or with both these parasites. The mites may probably be suppressed by a dressing of flour of sulphur, and other treatment has been recommended for destroying the fungus; but even if these recipes are effective, why should we submit to the importation of diseased goods? American farmers, gardeners and foresters have refused to do so, and the United States Government have adopted drastic measures for their protection, prohibiting the importation of plants and roots liable to certain diseases. This caused much inconvenience and loss to European exporters. At one time it was easier to land a European family in New York than a bag of potatoes; but the difficulty has been met by the establishment of a phyto-pathological service in some of the exporting countries, and many plants—including potatoes—are now admitted into the United States under a health certificate from the competent authority which, in Great Britain, is the Minister of Agriculture. Under

Flowers

the most recent relaxation of embargo, bulbs of *Galanthus, Scilla, Chionodoxa, Fritillaria, Muscari*, etc., are passed in without requiring a certificate. The importation of Dutch bulbs into this country was severely restricted for some years, but these are now admitted free if guaranteed sound by inspectors appointed by the Netherlands Government. Why should not similar regulations be enforced in respect of Japanese bulbs ? The trade must be of sufficient value to the growers to make them very unwilling to lose it ; their Government is certainly not behind that of other civilised countries in prophylactic enterprise, and it ought not to be difficult to convince them of the expediency of establishing a phyto-pathological service. But the initiative rests with ourselves. We are the sufferers, and so long as we are content to receive consignments of diseased plants, we have but ourselves to blame ; whereas if the importation of plants known to be vehicles of disease were prohibited except under certificate of health, it would be in the interest of all concerned—exporters and importers alike—that inspectors should be appointed to apply the necessary tests to the growing crops. This is a matter upon which it is most important that the Royal and other Horticultural Societies should take action by urging the Ministry of Agriculture to assume control over the trade in Japanese bulbs.

In planting lily bulbs it is essential to bear two points in mind—(1) Does the species relish lime, detest it, or is it indifferent to it ? and (2) Does it throw out basal roots only, or does it also throw out annual

PLATE II

⅓ natural size   LILIUM REGALE   2 August, 1919

Some Hardy Bulbs

roots from the flowering stem above the bulb, demanding in consequence to be more deeply planted ? It is not much use mentioning these two points without explaining how they are to be determined; but to do so would carry us beyond the fair limits of this chapter. The best service I can render to any of my readers who aspire to success with lilies is to refer them to the handbook on the genus contributed by Mr. A. Grove to the *Present Day Gardening* series, which can be had for half-a-crown, and contains the condensed precepts of one who has had more practical experience and success in cultivating lilies than any other person—professional or amateur—with whom I am acquainted. But all the precepts of all the experts are apt to be upset through the caprice of this class of plant, which determines them, at times, to reject the soil and situation most carefully prepared for them, and occasionally to attain extraordinary vigour and permanence in very unlikely places. Meanwhile, as I have experienced more failures with lilies than can have fallen to the lot of many amateurs, it may be useful if I name a baker's dozen with which I have succeeded, and which ought to fare satisfactorily anywhere with ordinary attention to their respective peculiarities :

1. The Giant Lily, *L. giganteum.* A woodland dweller; likes cool, deep soil and partial shade; detests lime, but relishes rich manure. Should be planted with the tip of the bulb flush with or just under the surface.

2. *Lilium monadelphum.*—A grand lily; likes lime, but thrives without it; intensely resents disturbance.

Flowers

Those now in the British market seem to be all of the straw-coloured variety, *Szovitzianum*. The canary-coloured form is taller and finer, but is now hard to come by.

3. The Nankin Lily—*L. testaceum*. Likes lime, but thrives without it. One of the loveliest lilies.

4. The Orange Lily—*L. croceum*. A most good-natured species; should thrive anywhere; revels in sunshine and stiff soil. Imported Dutch bulbs are not always true to type.

5. The Royal Lily—*L. regale* (Plate II). Plant in deep loam with lime, screened from burning sunshine. Seeds freely, but should only be propagated from the most robust strain.

6. Mrs. Sargent's Lily—*L. Sargentiæ*. Unlike the royal lily, to which it bears a general resemblance, this fine plant hates lime and loves full sunshine. It is easily propagated from stem bulbils.

7. *Lilium Davuricum* or *umbellatum*. Has sported into many forms, the best varieties known to me being *Incomparabile*, with chalices of gorgeous crimson and gold, and *Davuricum luteum*, of lowlier stature and flowers of clear apricot yellow. A good doer, either with lime or without.

8. The Martagon Lily—*L. martagon*. Likes lime and stiff loam. Developes best in partial shade or a north aspect. The white-flowered variety is a lovely thing, sending up a column of carved ivory Turk's caps with gold anthers. We raised a great quantity of it from seed, about 75 per cent. coming true. At the moment of writing there stands one near my

Some Hardy Bulbs

window five feet four inches high (not fasciated), bearing sixty-two ivory-white blossoms. The ordinary purple-flowered form near it is two inches taller. In the variety *Dalmaticum*, which the late H. J. Elwes inclined to consider a distinct species, the flowers are deep maroon, sometimes verging towards sable.

9. The Turk's Cap—*L. pyrenaicum*. Easiest of all lilies in our climate, thriving alike in sunshine and shade, in border or grass, with lime or without. The orange scarlet form, often supplied for the more refined and brilliant *L. pomponium*, is a gay and early flower.

10. The Tiger Lily—*L. Tigrinum.*—Dislikes lime, demands deep loam and moderate shade. The varieties *Fortunei* and *Splendens* are finer than the type; but if I had the chance I would set my heel on the last bulb of the double-flowered monstrosity. The special beauty of the Lily family consists as much in graceful form as in vivid or delicate colouring.

11. The Panther Lily—*L. pardalinum*. Thrives in loam and peat without lime. Although classed as a swamp lily in North America, it does not relish boggy soil in our more humid climate.

12. *Lilium superbum*. Enjoys similar conditions to those for the panther lily. It is the tallest of the genus except *L. giganteum*.

13. The Scarlet Martagon—*L. Chalcedonicum*. A brilliant flower, enjoying full sunshine and demanding thorough drainage. Like *L. monadelphum* it takes a couple of seasons to recover after transplanting, but is a hardy doer if it has the luck to escape *Botrytis*,

Flowers

from which deadly fungus I have found it more liable to suffer than any other species except *L. candidum*.[1]

After the amateur has succeeded with these, there are many other fine lilies awaiting his attention. In conclusion, let me mention two lessons borne in upon me by a long course of years chequered by success and failure—especially failure. First, that the best stimulant for all lilies is wood ash, provided it has been stored in a dry place, because the virtue thereof consists in the potash it contains, which a single shower of rain suffices to dissolve and wash to waste. And note that the ash of small stuff—brush, twigs, leaves and bracken—contains a larger percentage of potash than that of logs and large branches. Second, once any species of lily is well established and thriving happily, let no consideration in the heavens above or on the earth beneath or in the water under the earth induce one to disturb or move it. We received a sharp lesson to that effect in dealing with *L. pomponium*. This brilliant lily—the "martagon pompony" of Parkinson—appears to have vanished from the British market. Nurserymen's lists may be scanned

---

[1] I find that I have made no provision in my baker's dozen for *L. auratum* and *speciosum*, which should by no means have been omitted. Precepts for their treatment are admirably given in Mr. Grove's handbook. The Madonna lily also, *L. candidum*, is the only plant I know whereof it may be said that it ought to be in every garden. And so it would be, no doubt, if everybody knew and took the one precaution necessary to ensure it against *Botrytis*. That precaution is to keep it in such vigour as shall enable it to resist the invasion of that fell fungus. Shallow planting in good loam, with an annual allowance of sweet wood ashes, is the treatment which we have found successful.

PLATE III

⅛ natural size   LILIUM BROWNI   1 August, 1922

Some Hardy Bulbs

in vain for it; or if it happens to be offered, it is only too likely that you will receive the red-flowering variety of *L. pyrenaicum*, a vastly inferior plant. The reason for this is obscure, for it is said that the lily is still fairly abundant among its native Maritime Alps and it ought not to be difficult to obtain a fresh supply. Well, we had a fine clump of the true thing here growing in a poor stoney border full of the roots of an evergreen oak. The position seemed to the lily's liking, and it may be noted in passing that many species of lily are never so happy as among tree roots.[1] They are safe there from mice and grubs, and manage to draw enough nourishment to make good bulbs. But a bush of *Fatsia Japonica* grew to a great size, spreading its great leaves over the lily and shutting out the sunshine which it loves. Instead of sacrificing the *Fatsia*, I moved the pompon bulbs, ass that I was! and they have never made a decent show since.

It is a fascinating pastime to watch the leisurely manner in which lilies of the *Browni* and *Regale* class open their blossoms. Panthers and tigers, martagons and other Turk's caps seem in such a hurry to enjoy sunshine that they behave more precipitately than consists with dignity, cocking up a petal here, another there, before the rest are ready to move. Not so the great trumpet lilies. Slowly, very slowly the chocolate cylinders of *L. Browni* (Plate III), rise from a pendent to a horizontal posture; imperceptibly they lengthen and thicken, until one morning you shall see a chink of

---

[1] This does not apply to *L. giganteum, superbum, Henryi* or *pardalinum*.

Flowers

purest ivory open along the side of a five-inch bud. Next day the tips will have opened shyly; but it is not before the third day that they curl outward and backward, revealing the rich russet of the anthers and allowing delicious incense to be diffused.

Even a brief survey of bulbous plants such as this, wherein only a small fraction of desirable species have been noticed, would be culpably deficient if no mention were made of the loveliest of all South African bulbs —*Nerine Bowdeni*. It is the only one of a considerable genus that can be pronounced hardy under a south wall, and ranks among the very best wherever *Amaryllis belladonna* can be grown. It produces its large, bright rose blossoms with delicately crimped petals in September.

These pages were going through the press before I received a lesson worth recording about the treatment of tiger lilies—*L. tigrinum*. Many amateurs have complained about want of success with this fine plant, which, though it thrives permanently in some gardens, tends to dwindle and disappear from others. The fact is, it dislikes old garden soil; but when planted out in woodland, as I saw it recently at Logan in Wigtownshire, it rises to a height of over six feet and proves absolutely permanent.

II

The Herbaceous Border

> The daughters of the year,
> One after one, through that still garden passed;
> Each garlanded with her peculiar flower
> Danced into light and died into the shade.—*Tennyson.*

STANDING one fine September afternoon in a remarkable garden flanking a great castle in the north of Scotland, methought that it displayed the utmost opulence of colour attainable with hardy herbaceous plants. Down the centre of a spacious walled enclosure wound a sparkling rivulet, on either side of which the ground rose in smoothly mown slopes crowned by magnificent breadths of flower border, planted with great blocks of aster, antirrhinum, penstemon, helianthus, galtonia, wolfsbane, etc. One group consisted of thirty or forty plants of *Thalictrum dipterocarpum*, four and five feet high, choicest of all meadow-rues, unknown to European gardeners till it was introduced from China some twenty years ago. The pictorial effect of this flower-glen was splendid, and as I gazed upon it the thought crossed my mind how spotty and scrappy my own jumble of shrubs and herbs must seem to any visitor.

Flowers

It was a masterpiece of horticulture in that manner, yet there was something missing in it. Comparison may be odious, but it is inevitable; nor could I help comparing the impression received in this spacious demesne with that derived from a prowl—say round the far narrower bounds of Bitton vicarage garden in the late Canon Ellacombe's time. I may be told, perhaps with truth, that botanical pedantry was at the source of the greater satisfaction experienced in the smaller garden; and, certainly *not* with truth, that if Canon Ellacombe had owned ample pleasure-grounds and corresponding means to maintain them, he would have devised and achieved a similar scenic effect. It would have been contrary to his whole principles and practice to do so, because such an effect can only be accomplished by concentrated effort for display at a particular season, and sacrificing all beauty and interest in the borders for the rest of the year.

That was the rock whereon the mid-Victorian system of bedding-out was wrecked, enabling Mr. William Robinson to convince the owners of country houses of the unwisdom of laying their gardens bare and dismal for nine months in the year in order to have a floral pyrotechnic display during the other three. That, also, constituted the objection urged against the advice which I gave to the gardener in charge of another great pleasaunce in the north. The lady of the castle having asked me to take counsel with the said gardener about improvements, I perambulated the grounds with him. The place was ideal for the growth of Asiatic rhododendrons; deep sea

The Herbaceous Border

was close at hand to secure mild winters; wooded hills rose round three sides of the park to ward off blighting winds. I strongly recommended my companion to take advantage of such favourable environment, but he ruled me out at once. "The wur-r-rst of rhododendrons is that they'll not flower when the family's at home." And whereas the family is at home only when the London season is over and the shooting season begins, care must be bestowed only upon such flowering plants as are so tactful as to conform to the calendar of fashion.

That, then, is the motive in the gorgeous display in the garden first mentioned in this chapter. Plants have been chosen and cultivated with the utmost skill to be at their best " when the family's at home "— that is, in the autumn. One cannot have it both ways. Planting must be organised either for a full glut of colour at a prescribed season, sinking detail in general effect, or for such processional grouping as will provide against the borders being devoid of interest and beauty at any season of the year.

To one who loves flowers for their own sake there are few scenes more discouraging than a formal garden in the season when " the family's *not* at home." Such was my experience when, one April morning, I stood for the first time among the Italian parterres designed for Diane de Poictiers at languorous Chenonceaux. The place was then owned by an American gentleman (I believe it has since changed hands), who had scrupulously kept everything within and without the chateau in harmony with the spirit of the renaissance to which

Flowers

this fantastic dwelling and demesne owe their origin. Except in the garden, which had been adapted for bedding out, no doubt before the American had entered into possession. I had anticipated something reminiscent of the scenes which Brantôme has so faithfully recorded; of the philandering between the Dauphin and the lovely but coldly-calculating Diane; of Henri's formidable mother and her *escadron volant*; and surely there would be some tree or shrub to remind one of Mary Stuart, Queen of France for eighteen months, during which she was much at Chenonceaux, in the same manner that the mulberry trees at Hatfield revive memories of her kinswoman, Elizabeth.

> Elle était de ce monde où les plus belles choses
>   Ont le pire destin ;
> Et, rose, elle a vécu se que vivent les roses—
>   L'espace d'un matin.

But there was nothing. In all that flat, parcelled space of bare soil and pavement there was not a bush to screen a linnet—not a flower to blink in the sunshine, still less to waken association with the past, although on this brave April morn the meadows beside the smooth-flowing Cher were nodding with purple fritillaries. All growth and brightness were repressed until the season when, under decree of fashion, the lord of that fair manor and his household should spend a month or two there.

Very different was the impression received on first visiting the Elizabethan manor of Levens in Westmorland. Not only was the house but little changed since it was built round the old border peel which still

The Herbaceous Border

dominates it, but the garden wherein it stands remained faithful to the design which Monsieur Beaumont gave to it by instruction of that fiery Jacobite, Colonel Graham, who became possessed of the property in 1690. Beaumont was the Frenchman who laid out the garden at Hampton Court for James II, and although his fondness for topiary work is little to our taste at this day, Levens garden remains, or did remain at the time I speak of, a perfect picture of a country house pleasaunce in the stormy seventeenth century. Yews, preposterously clipped into similitude of everything least resembling their natural growth, stood in formal array along the paths; borders, fitted into an elaborate mosaic plan, had their formality mitigated by box edging half-a-yard high, within which a wealth of old-world flowers kindled each summer into flames of crimson, and white, purple and yellow, scarlet and blue; some of them, perhaps, the very roots which Beaumont caused to be set there more than two hundred years ago, for no man can assign a limit to the age of some herbaceous plants. Nay, it is possible that some of these herbs may have survived from the Elizabethan garden which was there before Beaumont took the ground in hand, when the estate belonged to the Bellinghams. Of the bowls which still remained on the ancient green, some were engraved with the Bellingham crest, others with that of Graham. The former may possibly have been turned by the same hand as fashioned the set with which Drake was playing on Plymouth Hoe when the report reached him that the Spanish Armada was off the Lizard.

Flowers

I have written of the garden at Levens as I used to know and love it; but many strokes of fate have fallen since those days. The kindly squire is no more; he died in the year before the great war, and his only son followed him seven years later, leaving the succession open to his uncle, Richard Bagot the novelist. Death duties and war taxation left scant margin for upkeep of pleasure grounds, and when I was last at Levens, the ancient parterres with their heavy box edgings and wealth of flowers had been levelled away, the ground being sown down with grass, leaving the yews still clipped into the semblance of lions, peacocks and apes in sombre melancholy, bereft of their old setting of bright colours. A few months after my flying visit, Mr. Richard Bagot breathed his last.

> We pass; the path that each man trod
> Is dim, or will be dim, with weeds:
> What fame is left for human deeds
> In endless age? It rests with God.

Let me not be suspected of under-rating the need for design in the disposal of plants in a mixed border. It is as essential as it used to be in bedding out, and far more difficult to carry out—partly because of the immensely greater variety of material—partly because, in order to secure a natural effect, design, which was always obvious in bedding out, must not be apparent. In dealing with hardy plants a cardinal precept is *Ars est celare artem*—art consists in disguising art.

These thoughts bring to mind another large garden —this time in the south of Scotland. The lady of the manor, an enthusiastic gardener, invited me to inspect

The Herbaceous Border

her herbaceous border, which was laid as broad and straight as a Roman road, about 150 yards in length. It was well stocked with excellent stuff bearing evidence of generous cultivation, but the hand of the planter was too much in evidence. The whole border was a chain of sections, each one filled with exactly the same collection of herbs as those on each side of it. The herbs were excellent; no expense had been grudged in collecting them, and they were carefully gradated in height from back to front; but there was no hint of the spontaneous—no mystery—no surprise—nothing to invite exploration after passing two or three sections —no happy accident of harmony or contrast.

It may seem that accidental effect is not consistent with the design that I have described as essential. Nay, but the design should be of that plastic kind which admits and avails itself of accidents. As the seasons run their course, chance combinations are sure to occur—some of them exquisite, others intolerable. For instance, it was not until the autumn of 1921 that I became possessed of *Narcissus Johnstoni* " Queen of Spain "—the gift of one of my daughters. One has become " fed up " with varieties of daffodil, and disposed to doubt whether any improvement is possible on the common *N. pseudo-narcissus*; wherefore, our borders being untidily crowded, I popped these bulbs into the first vacant place I could find. This happened to be alongside of a breadth of *Aubretia Græca*.[1] The

---

[1] Sometimes distinguished as a species, but probably only a fine form of *A. deltoidea*, in my opinion superior in habit and delicate colour to any of the others. Attempts to produce a red flowering variety of rock-cress have resulted in nothing but disagreeable shades of impure crimson.

Flowers

bulb and the herb flowered simultaneously in the following April, and so delectably was the soft chrome of the daffodil set off against the pale lavender of the rock cress, that means have been adopted to confirm and extend the accidental alliance. This Queen of Spain daffodil is reputed to be a national hybrid (accident again) between *N. pseudo narcissus* and *triandrus*, originating in the mountains of Portugal. It is one of the very fairest of an interminable clan, the segments of the perianth being airily reflexed, both they and the trumpet being of the same peculiarly luminous hue.

In late summer of the same year another happy combination arrived by chance. A stray plant of the willow gentian, *G. asclepiadea*, had sprung up, self-sown, in the middle of a mass of Potentilla "Gibson's scarlet," and the arching sprays of rich blue made admirable contrast with the intense vermilion of the lowlier herb.

So much for happy accident; it is only fair to record one of singularly infelicitous character. Fresh lodgment having to be found for some bulbs of the Chilean *Hippeastrum* (*Habranthus*) *pratense*, a sunny position was found for them under a south wall, without giving heed to the fact that on that wall was trained an Austrian copper rose. The bulbs and the rose flowered simultaneously, presenting an intolerable clash of vivid hues. In autumn the *Hippeastrum* was taken up, and dumped this time in front of a bush of *Buddleia globosa*. Again the result was distressing, for the two things flowered in the same month of June, and here there occurred triple discord, the front of the

The Herbaceous Border

border having been planted with *Geum Borisii,* whereof the red-lead flowers competed most unpleasantly with the orange *Buddleia* and the vermilion *Hippeastrum.* Indeed, it is not a simple matter to provide suitable neighbours for this flaming Chilean Amaryllid, and it was by chance again that we hit upon the intense azure of *Cynoglossum nervosum* as a fine setting for it, backed by the white sprays of *Libertia grandiflora.* But whereas the said *Cynoglossum* does not provide a mass of colour, only sprays of ultramarine, a still better comrade for the *Hippeastrum* is the Chinese *Iris chrysographes,* which increases fast and sends up sheaves of deepest purple flowers, with a thin line of gold embroidered on the fall.

Before strolling round the borders one should always slip into a pocket a handy note-book wherein to jot—among other things—such chance combinations of merit, or the reverse, as are sure to present themselves in such a ramshackle collection as we dignify here by the title of a flower-garden. Unless impressions of form and colour setting are fixed by memoranda as soon as they are received, they are sure to have passed from memory before the time when they can be applied to use. Here, for instance, is a note scribbled on a page marked " October "—(it must have been early in that month) :

" A bevy of blue-purple *Crocus speciosus* pushing up through a silvery carpet of *Cerastium tomentosum.* A tangle of *Polygonum affine (Brunonis)* beside porcelain blue *Geranium Wallichianum.* A furnace of *Kniphofia aloides* in a sea of snowy *Anemone Japonica.* A crimson cascade of *Berberidopsis corallina* falling over *Crinum Powelli album* at the foot of a wall."

Flowers

Among these notes are recorded the rich effect of *Muscari conicum*, "Heavenly Blue" massed among bushes of the rosy-white *Rhododendron ciliatum*; of the pale-blue *Anemone Appenina* behind an edging of sulphur-coloured *Saxifraga apiculata*; of the scarlet *Tulipa præstans* set off by a drift of the pale form of hoop-petticoat—*Narcissus bulbocodium citrinus*;[1] of the purple spires of *Salvia nemorosa virgata* as a setting for the Nankin lily—*L. testaceum*; while the orange lily—*L. croceum*—never looks brighter than in company with the double-flowering *Geranium pratense*, which blooms a month later than the single form. *Quid plura*? the list might be extended almost indefinitely, yet one more must be added—a crowd of blood-red blossom of the annual *Papaver umbrosum* round one of Lemoine's hybrid *Philadelphus*, a low-growing variety in the cataracts of snowy bloom, bringing to mind the promise—"Though your sins be as scarlet, they shall be white as snow."

Sometimes one may note happy triple combinations, such as the cool pink blossoms of aster "St. Egwin" set against the sky-blue *Aconitum Wilsoni* (a Chinese wolfsbane) and the canary-yellow *Lepachys* (*Rudbeckia*) *pinnata*. Again, *Cimicifuga simplex*—latest and best of bugworts—never shows to better advantage than in company with a rich carmine Lobelia and the quaint brimstone-coloured flowers of *Kirengeshoma palmata*, the last-named being entitled to a place were it only for its handsome foliage.

[1] This variety of the hoop-petticoat daffodil spreads freely self-sown; whereas we have found the golden-flowered type somewhat difficult to keep.

The Herbaceous Border

Terrestrial orchids form a class of garden plants which are too seldom seen in gardens, probably because their cultivation is beset with some peculiar difficulties. A few enthusiasts, however, have made a study of them and have shown that these difficulties are far from being inseparable. They are not fit subjects for the mixed border, as I have discovered too late and to my cost, having lost far more than remain to me. Hardy orchids deserve a special border to themselves, with due variation of soil from moist to dry and from lime to lime-free portions. It is only under such conditions that they can be secure from invasion and obliteration by robust herbs. The oriental poppy is a terrible bandit in that respect, for it scatters its seeds far and wide, and the seedlings, after lying low all winter, rush into such exuberant growth that the mischief is done before it can be foreseen. "*Orchis, Ophrys, Serapias, Spiranthes*, etc., are seldom grown," wrote Mons. H. Correvon in the *Garden* for 16th June, 1923, "the reason simply being that people fail to treat these orchids as tuberous plants, and so transplant them directly from the wild into the garden without giving them time to mature their tubers." It is, however, quite possible to take up orchids even when in full bloom, provided special attention is given to their root structure   The roots of tuberous orchids are of various forms. In *Orchis maculata* it is palmate; so it is in *O. latifolia,* but not so regular or so deeply divided. In *O. pyramidalis* it is undivided and ovally rounded; in *Habenaria bifolia* it is carrot-shaped, tapering to a point. All these forms have this in

Flowers

common, that the tuber wastes away and disappears after the plant has flowered, a new tuber forming alongside of it. As the life of the plant depends upon the new tuber, care must be taken that it be not knocked off or injured in transplanting.

One hesitates to advocate the cultivation of British orchids, some species being extremely rare, and others very local in their distribution. There are fleeting, but fervid, fashions in horticulture, witness the prevailing passion for narcissus; and if attention were drawn to terrestrial orchids as an attractive class, the industry of collectors might be stimulated in a most undesirable degree.

Some years ago I happened to sit at dinner next a very high dignitary of the Church of England. When a lark pudding was handed round, I refused it as promptly as if it had been Trimalchio's dish of dormice served with honey and poppy-seed; but my right reverend neighbour helped himself liberally to it. I ventured to express regret that he should so condone the slaughter of song-birds for the table. Was it Nero, Domitian or Lucullus, I asked, for whom an *entrée* of nightingales' tongues was prepared? His reply surprised me, forasmuch as, if valid for excuse, it might serve to cover as many sins as charity. "I suppose," said he, "if I did not eat them, somebody else would!"

It may be under some such flimsy casuistry that those creatures crouch who prowl about our woods and lanes and watersides, uprooting whatever plant —the rarer the fitter for their purpose—that can be turned into cash. Even if—as is doubtful—such

The Herbaceous Border

persons are endowed with conscience, how easy it must be to stifle it by the plea—" If I did not bag this plant, somebody else would ! "

It is to this incessant sordid pilfering that we owe the imminent or actual extermination of some of our choicest British wildings. The late Reginald Farrer, himself a famous collector among the affluent flora of Western China, has left on tragic record the fate of the noblest of British orchids—*Cypripedium calceolus*. This fine plant, which I wish could have been presented life-size in Plate IV,[1] was fairly abundant fifty years ago in certain parts of Yorkshire and Durham, and nowhere else in the British Isles. Let Mr. Farrer tell in his own graphic style why it may now be sought for there in vain :

" The Arncliffe valley is a narrow mountain glen, with steep fells rising through woods on either side towards the great moorlands overhead. . . . Here, in these mountain copses, ever since the time of Withering, the Cypripedium has been known. And one old vicar kept careful watch over it, and went every year to pluck the flowers and so keep the plant safe ; for without the flower you might, if uninstructed, take the plant for a lily of the valley. Then one year he fell ill. The plant was allowed to blossom ; was discovered, and uprooted without mercy, and there was an end of it. And worse is to follow ; for a professor from the north—I will not unfold whether it were Edinburgh, or Glasgow, or Aberdeen, or none of these, that produced this monster of men—put a price on the head of the Cypripedium, and offered the inhabitants so much for every rooted plant they sent him. The valley accordingly was swept bare, and, until the patient plant was re-discovered last year [1907] there was nothing left to tell of the glen's ancient glory except one clump of the Cypripedium which, to keep it holy, had

---

[1] It grows eighteen inches high, and the flower measures 3½ inches between the tips of the expanded petals.

been removed to the vicarage garden, there to maintain, in a mournful but secure isolation, the bygone traditions of Arncliffe.... Accursed for evermore, into the lowest of the Eight Hot Hells, be all uprooters of rarities, from professors downwards!"[1]

This lady's-slipper grows freely in many forests of Central Europe, where it may be able to resist, for a while at least, the insatiable rapacity of collectors. But the stock is not inexhaustible, even on the Continent. The time must come when the supply will fail, wherefore it becomes almost a duty for lovers of hardy plants to get this kindly orchid well established in British gardens. The chief difficulty consists in getting plants that have not suffered in removal and transit.

"Too often," says Mr. A. D. Webster, "has it been the case that persons who have attempted the culture of these plants have, after a year's trial, given it up in disgust and concluded by branding terrestrial orchids generally as both wayward and unmanagable. This, in the majority of cases at least, may be attributed to badly rooted or injured specimens being got to start with, for it should be borne in mind that perhaps no other class of plant suffers so much from injury, either to root or stem, as orchids.... I have seen a consignment of the very *Cypripedium* under notice sent over from the continent, to be sold at a shilling a piece in this country, that I would hardly have paid carriage on, and I would be within due bounds in stating that, of the several hundreds included in the consignment, ninety per cent must have succumbed, either to the ruthless manner in which they had been torn from their native haunts or to carelessness in packing."[2]

Whereas it is nearly forty years since this warning was uttered, it may be conceived that the prospect has not improved in the interval.

[1] *My Rock Garden*, pp. 31-34.
[2] *British Orchids*, by A. D. Webster, 2nd. edit., 1898, p. 18.

PLATE IV

½ natural size   'LADY'S SLIPPER'   21 May, 1918
(*Cypripedium calceolus*)

The Herbaceous Border

The root system of the lady's-slipper differs from that of most British terrestrial orchids in not being tuberous, but consisting of a matted tangle of long, fleshy roots. If these can be secured uninjured and without getting dry, the plant behaves most amiably in cultivation, asking only for deep, free loam with a dash of lime in it, and, being a woodland dweller, partial shade to screen off scorching sun. Never have I made a more satisfactory bargain than when, forty years ago or thereby, I bought from one who was quitting his garden in Surrey his whole stock—a dozen strong clumps—of lady's-slipper. They arrived fresh and moist; since which no month of May has failed to bring up the fine, bold leafage, surmounted by flights of maroon and yellow flowers.

So far as I have noticed, this lady's-slipper has never formed, still less ripened, seed in our border. Possibly the appropriate kind of insect visitor has not found the plants, or, when found, has failed to make its way to the nectary, approach to which requires a clue. The late Lord Avebury described the complex mechanism of the reproductive organs in this flower.

"The opening into the slipper is small, being partly closed by the stigma and the shield-like body that lies between the other two anthers, the result is that the opening into the slipper has a horse-shoe-like form, and that bees or other insects which have once entered the slipper have some difficulty in getting out again.... The pollen of *Cypripedium* is immersed in a viscid fluid, by means of which it adheres first to the insect and secondly to the stigma; whereas in most orchids it is the stigma which is viscid."[1]

[1] *British Flowering Plants*, pp. 412-13.

Flowers

The failure of this species to produce seed in our garden is singularly at variance with the prolific energy of most terrestrial orchids. Darwin reckoned the seeds in a single flower-spike of *Orchis mascula* at over 186,000, and calculated that if all these were to germinate and become seed-bearers, the third generation of their descendants would suffice to cover the whole surface of the globe with a green carpet.

After the lady's-slipper the showiest British orchid is *Orchis maculata*, especially the variety *superba*, popularly known as the Kilmarnock orchis,[1] which, grown in moderately-moist loam, without lime, increases to a fine clump eighteen inches high. Of our other indigenous orchids, some, like *O. latifolia*, the lime-loving *O. pyramidalis* and the ivory-white butterfly orchis (*Habenaria bifolia*) may be grown in nooks of the rockery; but their modest charms are sure to be overlooked in a mixed border. On the other hand, the Madeira orchis (*O. foliosa*) may be assigned rank with the choicest herbaceous plants, flourishing as free and fair under our watery skies as in its native sunny island; nor does it flinch, even from severe frost.

For five-and-twenty years and more we have been striving to get the queen of hardy lady's-slippers established here—the peerless North American *Cypripedium spectabile*. Established, I say, in the sense

---

[1] Said to have been found in a cottage garden near Kilmarnock by the late Miss Frances Hope, one of Mr. W. Robinson's earliest disciples in the revolution against bedding out. It was in her delightful garden at Wardie Lodge, Edinburgh, that, in the early 'seventies, I first gained insight into the cultivation of hardy plants.

The Herbaceous Border

that the lime-loving *C. calceolus* has become established in this lime-free district, pushing strongly every spring without any care or coddling. At first, following the prescription in gardening books, no doubt we erred in giving *L. spectabile* a boggy soil. I have never seen it in its native woods, but I understand that it flourishes in forest soil—that is, in deep, cool humus. Even if that be moist enough to be roughly described as boggy, the atmosphere of all districts of North America, except parts of the Northern Pacific coast is much less heavily charged with moisture both in winter and summer than that of the British Isles; and I fancy what *C. spectabile* dislikes is wet feet in winter. Mr. J. Stormonth, from whose nursery at Kirkbride, near Carlisle, I have had many good plants and useful notes about others, told me that it flourished best with him in almost pure sand. We treated some good clumps which he sent us in that way; but the result was no more encouraging than in previous years. They shot up fine fat spikes and flowered famously in the first season after planting, but dwindled away to nothing in after years. *Trillium grandiflorum* does so vigorously here, both in the borders and in the woods, that we should have supposed similar conditions would have suited the orchid; but that is far from being our experience. I feel all the more at a loss to devise treatment congenial to the wants of this fine flower, in that I have never happened to meet with it *permanently* vigorous in any British garden.

He or she who would derive full enjoyment of a garden should never be without a strong pocket lens,

Flowers

which reveals beauty that otherwise remains a monopoly for insects that visit the flowers. I happened to-day to examine for the first time a blossom of *Saxifraga aizoon rosea*, a pretty little species of the encrusted series, with panicles of flower that appear of a general pink tinge. Under the lens one of these flowers becomes a marvel of texture and colour. The petals are shown to be of brightly sparkling white, thickly spotted with little warts of intense crimson. The ten stamens of waxy consistency surround the ovary which shines like a dish of honey, the group of stigmata rising from the middle of it. Next I took a blossom of *Primula Japonica*, whereof the colour became of almost startling intensity under the glass, the rich crimsom of the petals merging into scarlet towards the centre, and then into maroon, both scarlet and maroon being sprinkled with little golden glands which altogether escape the naked eye. The interior of the tube shows like a glowing orange well. I need not prose further about this matter; but I am convinced that those who care for flowers will never realise their full beauty till they apply a lens to them.

It need hardly be said how greatly the appearance of a herbaceous border depends on the care bestowed in tying up or otherwise supporting such plants as flop after a heavy shower or are liable to be bent or broken by the wind. Larkspurs, *Romneya Coulteri* and some other things of bushy habit are best suited with pease-sticks, which, if put round the plants early in May, soon cease to offend the eye under cover of the growing leaves. In staking other plants that

The Herbaceous Border

cannot remain erect without aid, the bast or raffia usually employed is often exasperatingly weak; even if fairly strong at first, it soon rots in wet weather. To those who grow New Zealand flax—*Phormium tenax*—or, failing that, can beg, borrow or steal a sheaf of the great leaves from a friend who does grow it, let me recommend it as an invaluable substitute for bast. Leaves five or six feet long divide easily into serviceable strips of extraordinary strength, the only care to be observed in using them is to avoid tying "granny" knots, for the fibre is stiffer than bast and requires to be firmly fastened. It is also most useful for tying up parcels of plants and other things for transit by post or otherwise.

"The point where the garden ends," remarks a recent writer on laying out grounds, "and the woodland begins should be a place of joy. Trees are great kindly friends; they extend welcoming arms and utter welcoming sounds."[1] True enough, but trees are wont to extend roots as well as arms, and the sounds which the owner of the garden utters when he finds his borders full of these roots may be the very reverse of "welcoming." Horse chestnut, elm, ash and poplar are the worst neighbours in this respect: it is astonishing to what distance they will send their roots to suck up nutriment. In our modest flower-garden at Monreith the subsoil is rock in parts and glacial drift, *i.e.* boulder clay, in others, through neither of which can trees drive their roots. It has been, therefore, a

[1] *Gardening for the Twentieth Century*, by Charles Eley; London: John Murray, 1923.

Flowers

fairly simple operation to circumscribe trees within the enclosure with a barrier of concrete resting on the subsoil and brought flush with the surface. Even so, it requires vigilance to detect frequent attempts to surmount or burrow under the barrier, for trees of the kinds aforesaid seem to possess a sense enabling them to detect the presence of rich soil a long way off. In gardens where the soil is deep and permeable by roots it is impossible to fence them off subterraneously; the only safe course is not to grow trees of that character within or close to the garden. Conifers, holly, evergreen and deciduous oaks, hawthorn and beech, do not give much trouble as robbers.

Those whose lot is cast on an exposed sea-board, as ours is at Monreith, must rely greatly for shelter upon tall hedges which have to be closely clipped to prevent them getting bare below. The tree usually recommended, and most commonly planted, to form a garden hedge is the yew, a tree which has to reach a great age before it is worth anything except as a foil to more lightsome growths, and one whereof the swarthy foliage should suffice to exclude it from any place near a dwelling-house. I am aware that this will be accounted rank heresy by many garden lovers, but I claim to have the poets on my side. Shakespeare speaks of the " dismal yew,"[1] and Sir Walter Scott applied the same epithet to it :

> But here 'twixt rock and river grew
> A dismal grove of sable yew;
>
> . . . . .

[1] *Titus Andronicus*, ii. 3.

The Herbaceous Border

> Seem'd that the trees their shadows cast
> The earth that nourished them to blast;
> For never knew that swarthy grove
> The verdant hues that fairies love;
> Nor wilding green, nor woodland flower,
> Arose within its baleful bower;
> The dark and sable earth receives
> Its only carpet from the leaves.[1]

There are many more cheerful evergreens than yew, and equally effective as wind-screen. Even beech or hornbeam, though deciduous, exchange their summer verdure for a rich brown mantle which they retain throughout the winter when clipped as a hedge. And who may dissent from John Evelyn's praise of the holly, emphasised, as was his wont, by significant typography.

"Above all the natural *Greens* which inrich our *home-born* store, there is none certainly to be compar'd to the *Agrifolium* (or *Acuifolium* rather) our *Holly*, insomuch as I have often wonder'd at our *curiosity* after foreign Plants and expensive *difficulties*, to the neglect of the culture of this *vulgar*, but *incomparable* tree.... Is there under *Heaven* a more glorious and refreshing object of the kind than an impregnable *Hedge of near three hundred foot in length, nine foot high and five in diameter*; which I can shew in my poor *Gardens* at any time of the year, glitt'ring with its arm'd and vernish'd *leaves*? the taller *Standards* at orderly distances, blushing with their natural *Coral*. It mocks at the rudest assaults of the *Weather*, *Beasts* or *Hedgebreakers*."[2]

There speaks the whole-hearted amateur, and we

---

[1] *Rokeby*, canto ii., 9. The fidelity of this picture will be recognised by any visitor to the only woods consisting only of yews that are known to me, namely, the Great Yews and the Little Yews on Lord Radnor's property near Salisbury. The gloom of these singular groves is oppressive on the fairest summer day.

[2] *Sylva, a Discourse of Forest-Trees*, 3rd edit., 1679, cap. xxvi.

Flowers

at the distance of more than two centuries, cannot withhold our sympathy from him in the grievous injury wrought upon his hedge by the Czar Peter the Great, to whom he leased Sayes Court in 1697. For that most energetic monarch sought relaxation from the manual labour he imposed upon himself in Deptford dockyard by causing his courtiers to trundle each other in wheelbarrows down a steep slope into the holly hedge, whereby it was sorely broken and defaced.

Several kinds of conifer make beautiful hedges; in the west country we find nothing so excellent for that purpose as the Monterey cypress—*Cupressus macrocarpa*—which grows very rapidly from cuttings, stands clipping well and is always of a lively green. It is eminently a maritime species, suffering from frost inland, where Lawson cypress—*Cupressus Lawsoniana*, Nootka cypress—*C. Nootkatensis* or *Thuja plicata* may be used instead.

Except in forming and keeping hedges, the less use of shears in the flower garden the better. The present writer is one of a fast-dwindling band of amateurs who can recall the opening of Mr. William Robinson's spirited crusade in the early 'seventies against mid-Victorian gardening, and who rejoiced with him in its victorious result. It led to the utter discomfiture of the bedding-out school, restored to our borders the old-fashioned perennials which still lingered in the kitchen gardens of country houses and in cottage plots, and whetted our appetite for the innumerable new species that have been landed on our shores in recent years. It is good to find Mr.

The Herbaceous Border

Robinson faithful to his creed in his latest book,[1] inveighing, with all his pristine energy, among other things, against the distortion of trees and shrubs by topiary work.

"As to the clipping folly," says he, "many good country houses are disfigured by it. One of our weekly papers devoted to country life has lately figured a place in Yorkshire with the trees shorn into the shape of old Jerry hats.... What gain is it to the noble art of architecture to flank and fortify a fine building with caricatures of trees?"[1]

The amount of clipping carried out in some demesnes that might be named not only produces a wearisome iteration and uninteresting effect, but absorbs a vast amount of labour that might be turned to more intelligent account.

The question is often argued whether annuals should be admitted to the company of herbaceous plants. The only sensible answer takes the form of another question—Why not? A negative could only be founded upon some obscure point of horticultural etiquette, not worth considering. Some of the sweetest and most brilliant flowers are only annual or biennial, and no reasonable objection can be urged against giving them a place among perennials. They have been discredited by the manner in which they are too often treated. The seed is sown thickly in patches; little attention, or none whatever, is paid to thinning the seedlings, which grow up a dense, etiolated throng of weaklings, producing but a feeble caricature of what the individual plants can do if they but get a chance.

[1] *Home Landscapes*, 2nd edit. London, John Murray, 1920.

Flowers

One September day I was fishing the Spey, my beat being the lowest on the river where it flows into the sea through a vast expanse of shingle, almost bare of vegetation. Noticing afar off a gleaming patch of orange, I toiled in my wading breeks across the banks to find out what it was. It turned out to be a single plant of *Eschscholtzia Californica*, chance sown, a yard wide and one mass of fiery orange. I realised for the first time of what this plant is capable when it is given room, and I laid the lesson to heart, applying it thenceforward to the treatment of Shirley poppies, corn flowers, *Godetia*, mignonette, *Phacelia campanularia* and other annuals. If anything is worth doing, it is worth doing *well*. No farmer would expect a decent crop of roots if he neglected to thin seedling turnips in the drills; every kind of wild plant has to fight for its place in the sun, a life-long mortal struggle from which to rescue it is the gardener's cardinal function. Let any one try the experiment of sowing two plots A and B, each six foot square, with—say—Shirley poppy. Thin out the seedlings on plot A so that each shall have a space to itself four inches square, and thin out those on plot B so that each shall have a space one foot square. That will leave a crop of 108 poppies on plot A against 36 on plot B. The difference in the result may surprise him. Plot A, containing three times as many plants as plot B, will neither make half the display nor will it last half as long in flower, the vigour of the plants in plot B being many times greater than those in plot A.

I have chosen Shirley poppies to illustrate the

The Herbaceous Border

matter in print because, while most annuals may be easily transplanted when young, poppies do not agree with disturbance and have to be sown where they are meant to grow. In passing, let me pay a grateful tribute to the memory of the Rev. W. Wilks, to whose patient skill we owe this glorified form of the common corn poppy. Its only defect is that heavy rain beats down the plants into a hopeless tangle; but provision may be made against this disfigurement by setting small branches firmly among the seedlings to support them as they grow. The branches are soon hidden by the leaves and flowers. On the general question of the admission of annuals to the company of perennial plants, surely the objection on the score of propriety may be dismissed. Blanks will appear in herbaceous borders from time to time, and may be temporarily made good by putting in an annual or two, either transplanted from nursery beds or as thinnings from other parts of the garden.

Biennials may be treated in the same way, much of the rich effect of a mixed border would be missed if white foxgloves, evening primrose (*Oenothera*) and antirrhinums were ruled out; while for delicacy of colouring there are few flowers that equal the blue poppies of the Himalaya and Chinese mountains. It was only after many experiments that I hit upon a thin wash of Antwerp blue as the nearest approach to the exquisite hue of Bailey's Thibetan variety of *Meconopsis simplicifolia* (Plate V), which bears large, solitary flowers on eighteen inch stems, the golden anthers being carried on violet stamens. The

Flowers

commoner form of this species has dark blue flowers, tending in some varieties to purple. *M. latifolia* usually sends up spires of azure blossom to a height of two feet, but some of the seedlings produce pure white flowers and others white tinged with blue. It sows itself pretty freely here on a retaining wall facing west, and has a charming effect on the steep slope. *M. Wallichiana, Nepalensis* and *paniculata* grow to a height of four to six feet, with yellow, blue or purplish flowers; other species being *M. integrifolia,* sulphur-coloured, *M. punicea,* scarlet, and *M. Pratti,* dark blue with white anthers; but *M. simplicifolia* and *latifolia* are loveliest of the lot. All species produce abundance of seed and are perfectly hardy. They may be pricked out in boxes and planted out in autumn, care being had to label their position, for the leaves of some species—especially *M. simplicifolia*—disappear entirely in winter; but the finest plants are those which spring from seed self-sown or scattered by hand in suitable places and never disturbed. And if you want to present a *tableau vivant* of Beauty and the Beast, plant among them two or three tubers of the monstrous dragon's-mouth—*Helicodiceros (Arum) crinitus* (Plate VI)—which only asks for a sunny space where it may lay open its yawning, blood-stained spathes measuring a foot in length and nearly as much in breadth.

PLATE V

½ natural size

MECONOPSIS SIMPLICIFOLIA
(*Var. Baileyi*)

13 May, 1923

III

Some Flowering Shrubs

> Of a' the airts the wind can blaw
>   I dearly like the west;
>
> .   .   .   .   .
>
> Oh blaw ye westlin winds, blaw saft
>   Amang the leafy trees;
> Wi' balmy gale frae hill and dale
>   Bring hame the laden bees.
>   *Robert Burns.*

THAT the mean winter temperature of the west coast of Great Britain and a large part of Ireland is higher than that of the midland and eastern districts, is a matter of common knowledge to gardeners and amateurs. Probably most of them take the phenomenon for granted, without speculating as to the cause; but one often hears it attributed to the direct influence of the Gulf Stream. But in fact that great volume of warm water, which has a temperature of 65° Fahr. on leaving the Gulf of Mexico, ceases to be distinguishable as an ocean current east of 30° west longitude, 600 geographical miles distant from the coast of Ireland. The Gulf Stream, therefore, cannot be accounted as reaching the British isles as an

Flowers

operative current. Logs of timber, hard-shelled fruits and other flotsam thrown upon our shores no doubt may have been carried a considerable distance by the Gulf Stream, but owe the rest of their voyage to surface drift under the prevailing south-west wind. Nevertheless, the Gulf Stream materially affects the temperature of the Atlantic Ocean, conveying into it and distributing an immense volume of heat, which has been estimated to be equal to one-fifth of the heat received from the sun by that ocean between the Tropic of Cancer and the Arctic Circle. Indirectly, therefore the climate of our western seaboard, and in a less degree all parts of the British Isles derive advantage from the Gulf Stream, after it has ceased to be a stream or current through its agency in warming the general body of the Atlantic Ocean. More direct, however, is the effect of another agent, namely, the general movement of the atmosphere in our latitude from S.S.W. to N.N.E. This relatively warm current of air is heavily charged with moisture, which, in virtue of its own latent warmth, it carries in the invisible form of vapour. In winter the mean surface temperature on the ocean is higher than that of the land, whence it comes to pass that on meeting the land the S.W. air current receives a chill, causing the vapour which its winter temperature of 40° to 50° Fahr. has enabled it to carry to be condensed into mist or to be precipitated in the form of rain or snow. Moreover, on many parts of the west coast of Britain the land immediately facing the sea is of considerable height—in places even mountainous—whereby the atmospheric

Some Flowering Shrubs

current is suddenly raised to a higher and colder level. It has been calculated that if the said current were perfectly dry, its temperature would be lowered one degree Fahrenheit for every 182 feet of rise (one degree Centigrade for every 300 metres). But so far from being dry, it is saturated with moisture, which is condensed on meeting the chill of a higher elevation, and falls as rain or snow. The latent heat, thereby released from its burden, goes to raise the temperature of the district. It has been calculated by Dr. Haughton that " one gallon of rainfall gives out latent heat to melt 75 lbs. of ice or 45 lbs. of cast iron,"[1] and that on the west coast of Ireland the heat liberated by rainfall is equivalent to half the amount derived from the sun.[2]

For nearly two hundred years after the Roman occupation of Britain had come to an end and the last of the legions had been withdrawn in A.D. 410, there is a total absence of authentic record of events in that island and the conditions of its people. Procopius, indeed, writing somewhat more than one hundred years after that date, gives a detailed and extremely inaccurate description of Britain. He seems to have been patriotically anxious to make light of the loss by

[1] *Physical Geography*, p. 126.

[2] The science of meteorology has been greatly advanced since the theory upon which these remarks are founded was formulated ; but I believe that the cause and effects of orographic rainfall, especially in relation to the climate of our western seaboard, are still accepted in the main as stated above. To obtain a more comprehensive insight into the general circulation of the atmosphere recourse should be had to Sir Napier Shaw's recent work, *The Air and its Ways* (Cambridge University Press, 1923).

Flowers

the empire of the northern part, at least, of that province; for, while he admits that the southern districts, nearest to the continent, are fertile and well cultivated, he describes Caledonia as being ravaged by savage beasts, and cursed with a climate so poisonous that human beings could not inhabit it!

This statement, gravely stated in the Roman historian's account of the war with the Goths, came to mind when, several years ago, I was discussing with a well-known London publisher the projection of a volume on Scottish gardens, to be illustrated with reproductions of Miss Mary Wilson's masterly drawings in pastel. After looking through the drawings he asked me quite seriously—" Can you *really* grow flowers like these in Scotland ? "

I smiled at the implied slight upon our Scottish climate; but I had no right to do so, forasmuch as at that time we did not understand of what that climate was capable. *Sero sapiunt Phrygii*—the Phrygians are slow to learn—and there be many folks nearer home in like case. Cornish gardeners, I believe, were first enlightened about the superiority of their climate over that of the Midlands by Sir Joseph Hooker, who lavished upon them some of the treasures which he had brought from the Himalaya, and revealed to them that it was possible to cultivate in the extreme south-west of our island trees, shrubs and herbs which had hitherto been treated as subjects for a greenhouse. It is recorded of him that, when he revisited the Cornish *riviera* many years after the planting which he had recommended to be done, he

½ natural size   'DRAGON'S MOUTH'   30 June, 1920
(*Helicodiceros crinitus*)

PLATE VI

Some Flowering Shrubs

exclaimed—" Why, rhododendrons flourish better here than in Sikkim !"

That must have been in the latter half of the nineteenth century; but wellnigh fifty years were to pass before we realised that in many places on the west coast—in fact, wherever shelter can be provided against strong winds—conditions of soil and climate are quite suitable for many plants usually classed as half hardy. In preparing his great work on rhododendrons Mr. G. Millais visited almost every place where the finer species are grown, and came to a conclusion as follows :

"In Scotland, where conditions on the west coast are, I think, even better than in Cornwall, there are but few gardens devoted to rhododendrons. Mr. Osgood Mackenzie has a remarkable place at Poolewe, on the west coast of Ross-shire, where he can grow almost anything sub-tropical, and where certain Australian and New Zealand plants flourish with remarkable vigour. Western Argyll and Dumbartonshire should be one beautiful garden, as most moisture-loving shrubs find a congenial home there. . . . At Ardarroch (Gareloch-head) there are plants of *Rhododendron arboreum* up to 24 feet in height, and *R. campanulatum* seeds itself there as freely as in the Himalayas."[1]

It is a source of unavailing regret to us old folk that we did not find out the capabilities of our western climate long ago. Now that we have done so, some of us are preparing a sumptuous feast of form and colour for those who shall presently stand in our shoes. Far more persons are bestowing intelligent and discriminating care upon their gardens than was the case fifty years ago ; if they find any difficulty in stocking

[1] *Rhododendrons*, by J. G. Millais, pp. 70, 71.

Flowers

their borders it arises, not from want of good material, but, so vast is the number of new and attractive species introduced of late years from the temperate zones of both hemispheres, from sheer perplexity in making a good selection from the host of good things to be had at reasonable prices.

One of the gardening weeklies lately invited its readers to draw up a list of the twelve most desirable flowering shrubs. A considerable number of persons responded by naming the dozen which each of them preferred. The practical result was not more convincing than the late Lord Avebury's prescription of the hundred best books. The lists sent in revealed a great variety in the taste—or experience—of the writers. So much the better. If all had agreed in naming the same twelve shrubs, those twelve might have forthwith been planted by every reader of that journal; whereas one of the chief objects to be aimed at in furnishing a garden is that it shall not be a replica—either in design or contents—of any other garden. The intrinsic beauty of a plant is not enhanced by its rarity or novelty; some of the commonest kinds are among the loveliest, and no tree or shrub that has been brought to this country of late years outshines in splendour the horse-chestnut, the laburnum, the wistaria or the varieties of our native hawthorn. No park or pleasure-ground can afford to dispense with good specimens of these and other things that may be classed as common, and no private owner's collection can contain more than a fraction of the myriad desirable plants that have been acclimatised in this

Some Flowering Shrubs

country. The number is so vast that it may prove of some little use to any reader who contemplates beginning or extending a collection of hardy shrubs if I give my experience with a few of the choicest flowering shrubs that are less commonly grown than others. Being about to prose about rhododendrons in another chapter, I will leave that grand genus aside for the present; and before proceeding to business, let me relate how a famous diplomat gave me a lesson which the mention of hawthorn as an old favourite has brought to mind—the point of the lesson being that one ought not to let familiarity breed—I will not say contempt, but indifference

On a day in early May, 1913, the late Mr. Joseph Choate, so long and affectionately esteemed as American ambassador at St. James's, drove me in his car across Long Island to lunch with Theodore Roosevelt at Sagamore Hill. The woods through which lay our route were full of *Cornus florida*, loaded with white— on some bushes with bright rose-blossom.

" What a splendid thing that is," I exclaimed; " how I wish we could grow it decently in Britain ! "

" Well, yes," replied Choate, " it is a pretty plant; but I don't think it can compare with your English may."

Each of us inclined to prefer the flower with which we were less familiar.

" Our English may "—nor should patriot Scot or fiery Irishman demur to the limitation implied in the phrase, albeit that north of the Tweed it flourishes as fair and free as in any English county, and that the

Flowers

most remarkable assemblage of aged hawthorns that I can call to mind is in the Phœnix Park of Dublin. If fault there be, it lies with our Scottish bards, who have strangely neglected this beautiful tree, which Southron poets have garlanded with rhyme, usually under the name of May from its association with the May-day festival in honour of the Virgin Mary, which was universally observed by all classes—princes and peasants alike—until the Reformation laid a ban upon it.[1] Chaucer, however, uses the older name of hawthorn.[2]

> Thus sange they alle the service of the feste
> And that was done right early, to my dome,[3]
> And forth goeth all the court, both most and least,
> To fetch the flours fresh and branch and bloom,
> And namely [4] hauthorn brought both page and groom.[5]

The summer of 1922 was exceptionally cold and wet, but it was memorable above all others in my recollection for the extraordinary display of may-blossom, the result of abnormal heat and drought in 1921. That display was followed by an immense crop of haws, which surely should discredit, if anything could, the obstinate myth, so dear to the provincial press, that such a crop portends a hard winter—" Many haws, many snaws," as we say in the north, as if it were possible for a coming winter to affect growth in

---

[1] Hawthorn is never, or hardly ever, in bloom on May day according to our Gregorian calendar; but under the Old Style (Julian) the first of May corresponds with the 12th or 13th according to our reckoning—I am not sure which.

[2] Anglo Saxon *hægthorn*, alba spina.   [3] According to my judgment.
[4] Specially.   [5] *The Court of Love*, stanza 204.

Some Flowering Shrubs

a passing summer. The amount of fruit born in any season by the hawthorn and by most all other trees that flower on the previous year's growth, depends primarily on the character of the weather in the summer preceding that in which it is formed, affecting the degree in which the young wood of that season has been ripened.

It would fill a very large volume to treat of the multitude of fine-flowering shrubs or small trees that have been brought from the ends of the earth and successfully cultivated in this country. Mr. W. J. Bean has accomplished a great work of that nature to which gardeners and amateurs turn continually for sound information, with the certainty of never being put wrong. But whereas his two volumes contain a descriptive notice of upwards of four thousand species and varieties, it is not every one who has the leisure or experience to enable him to make choice among such a multitude of good things. It may prove of some small service if I draw attention to a few of those that have rewarded us best and grown most satisfactorily at Monreith. It will be understood, I hope, that what follows is far from being intended as more than a selection from plants that have prospered with us, with notes about a few of the many that have failed.

I need not dwell long upon European plants with which most of us are fairly familiar; but I would say a word about the laburnum, which indeed cannot fairly be reckoned as a shrub. My excuse is that I find that many persons who are passionately fond of flowers, but have not given practical attention to their

Flowers

production, have not realised that there are two distinct species of this splendid tree, and that by judiciously planting both the flowering season may be greatly prolonged. The common laburnum—*L. vulgare*—is the earlier to bloom, beginning in the first fortnight of May (sooner in the southern counties), and just as its golden tassels are flagging a fortnight or three weeks later *L. alpinum,* called by nurserymen the Scotch laburnum, takes up the running and continues till June is far sped. Its racemes are double the length of the other's and of a richer yellow—altogether a finer plant. The variety named *Watereri* carries even longer tassels. In choosing a place for it, due allowance should be made for its extremely rapid growth and far-branching habit.

The Judas tree—*Cercis siliquastrum*—requires more sun than can be provided for it in northern England and Scotland; in fact, I do not remember to have seen a decent specimen north of the Humber, except as trained on a wall, which destroys the character of it completely; but in the southern English counties it presents itself as a cloud of soft carmine in May, and those who inhabit such districts would be well advised to plant it more commonly than they do. The Cistus family are also sun-lovers, as is patent to anyone who has seen the marvellous profusion with which the different species clothe the most arid mountains in Southern Europe. There is no more hopeless task than an attempt to uproot even small seedlings from the hard-baked soil wherein they have their wirey roots. Often as I have tried it, never yet have I

Some Flowering Shrubs

succeeded in bringing one home to this country alive. Yet, strange to say, all the species which are grown here, seven or eight in number, make themselves perfectly at home, enduring our sloppy winters and tepid summers, growing and flowering luxuriantly. It is winter cold, not winter wet, that punishes them, or the combination of both. *C. laurifolius* is reputed as the hardiest of the family, that and the hybrids *C. Corbariensis* and *Loreti* being the only species which Mr. Bean certifies as having survived at Kew the memorable winter 1894-5. Nevertheless, there are many districts not much milder than the average of British climate where several other species may be grown. Even if old plants succumb to unusual cold, it is a simple matter to keep a young stock in reserve, for no plants are more readily raised from cuttings. Only yesterday I saw a fine specimen of *C. populifolius* in full flower on the rock garden in the Edinburgh Botanic Garden, where winter is the reverse of languorous. Here they give us no trouble at all ; the profuse gaiety of their flowers and the long period during which the succession is maintained render them among the prime favourites of border and shrubland. The handsomest of all is the true gum cistus—*C. ladaniferus*—with flowers more than four inches across, each of its white petals stained with a blotch of the colour of clotted blood ; but it is one of the more tender species. *C. Cyprius* is its offspring by alliance with *C. laurifolius,* and is much hardier. It often passes for the true gum cistus,[1] but is easily distin-

---

[1] It is wrongly figured for *C. ladaniferus* in the *Bot. Mag.*, t. 112.

guished from it by its smaller flowers borne in clusters, whereas those of *C. ladaniferus* are solitary. Mr. Bean observes that there is no cistus with yellow flowers though several species have a yellow stain at the base of each petal. No sooner, however, do the great petals of these two species fall, as they do in the afternoon of each day, than they turn bright sulphur-yellow. In Andalusia, where some of the hillsides are white for miles in the morning with the bloom of *C. ladaniferus*, they are brown and dark green by four o'clock afternoon, but under each bush is spread a yellow carpet of fallen petals. The same change may be observed in the flowers when pressed and dried.

C. purpureus, probably a natural hybrid between *C. villosus* and *laurifolius*, has larger flowers than any other of a rose colour, and is a truly brilliant thing. The epithet *purpureus* is misleading, for the petals are clear rose with a rich crimson blotch. In Mr. Hillier's nursery at Winchester there is what seems to be a spontaneous hybrid between *C. Cyprius* and *crispus*, which he has named " Silver Pink "—a pretty thing. Another hybrid or variety of *C. crispus* goes by the name of " Sunset "; its flowers are a fine crimson, but I know not where the variety originated. Some years ago a plant arrived here (whence I have forgotten) under the name of *C. recognitus*, bearing flowers in clusters, each of them three inches across with a remarkable flash at the base of the crinkled white petals, the inner half yellow, the outer bright carmine, pencilled with dark maroon lines. The effect of this arrangement round the central boss of orange anthers

Some Flowering Shrubs

is exceedingly rich. It is one of the prettiest of the genus, grows between four and five feet high and is said to be a natural hybrid between *C. laurifolius* and *Monspeliensis*. I have never happened to meet with it in any other garden and strongly commend it to the attention of those who can grow such plants in the open.

One approaches the family of *Cratægus* with some trepidation, for has not Professor Sargent recognised no fewer than one hundred and thirty-six species in North America alone! I shall content myself with drawing attention to one only, and that a European species, *C. Carrierei*, a plant of hybrid origin with corymbs of white flower in June, followed by large red and orange fruits, which hang on the branches through the winter. This is a most desirable small tree, the handsomest of all the family except our native hawthorn.

If North America had furnished us with no other shrub than her calico bush (*Kalmia latifolia*), British gardeners would owe her a debt not easily repaid, for it beats me to name a hardy, flowering evergreen excelling it in delicate beauty. Moreover, unlike many shrubs from the Eastern United States, it makes itself perfectly at home on this side of the Atlantic, and there is nothing to hinder any of our people enjoying a display thereof such as June produces in the grounds of the Arnold Arboretum. We have only ourselves to blame for neglecting it as we have done, for it is quite a rare treat to meet with it well-grown in this country. It is not more exacting in its requirements than any of

Flowers

the commoner hybrid rhododendrons with which our parks and gardens have been crammed. It is worth while to examine the flower of this charming shrub, which is equipped with a mechanism as effective for securing cross-fertilisation as that to be noticed presently (page 73) in the barberry family, but on a totally different plan. At the back of each of the five lobes of the corolla there is a little knob or cusp which not only serves to enrich the appearance of the truss, but fulfils an important function in the economy of the plant. When the flower expands, the brown anthers are fixed in the cups corresponding to the knobs at the back of the corolla; and the act of expanding bends the white stamens outwards. On a sunny day when the flower is dry, a slight touch with a pin placed under the stamen releases the anther, and the stamen springs up, flinging the ripe pollen to a considerable distance.

Before the year is six weeks old I am prone to speculate what must have been the feelings of him for whom *Ribes sanguineum* first hung forth its rose-red tassels in this country. It is no novelty in the eyes of the present generation, for one hundred years have nearly run their course since it was introduced by David Douglas, who, in the early nineteenth century, played for us in North America the part that E. P. Wilson, Reginald Farrer, Forrest, Kingdon Ward and other intrepid collectors have done and are doing in China, but it should still be prized as one of the very choicest of woodland undergrowths, ripening plenty of its indigo berries and springing up freely wherever the seeds fall in a vacant space. Moreover, it is possessed

Some Flowering Shrubs

of some property which deters the rabbit—of all beasts of the field the most omnivorous vegetarian—to exclude it from its dietary. No other currant will stand comparison with it for profusion of beauty, and no degree of frost whereof the British climate is capable prevails to affect its constitution, although a cold spell may retard its bloom. According to a record kept here of the kind which, if you would be accounted learned in botany you must speak of as " phenological," that is, pertaining to the leafing and flowering of plants—the date of the earliest blossoms on this ribes has ranged between 21st January and 31st March. Unlike the hawthorn and most other flowering shrubs, hard clipping does not interfere with its display, and a hedge thereof becomes sheeted with crimson—a remarkable sight in spring.

Ribes Gordonianus is the offspring of an illicit alliance between the crimson *R. sanguineum* and the yellow *R. aureum,* with their colours prettily combined in the flowers, but not so bright as in either parent. Many years ago I got a plant under this name from a nurseryman of good repute, but after I had waited several seasons, it turned out to be a worthless thing with dingy green flowers. It has now grown so big that labour cannot be spared for its removal. The same tradesman supplied me with another species of currant which he certified as the rare *R. Missouriensis.* I condoned its sprawling, weakly habit in consideration for its bright autumnal colour; but whereas Professor Sargent has since identified the plant as the common *R. Americanum,* I feel half inclined to

Flowers

avenge myself for this double injustice by giving the name of the offending firm. Howbeit, I refrain, having received too many good things from that source in the past. Moreover, the founder and head of that firm is now in the fields of asphodel, where complaints from discontented customers may vex him never more. *Requiescat!*

Mistakes of this kind may easily occur in a large nursery; the wonder is that they are not more frequent. None the less are they peculiarly provoking, for one may have to wait many years before the error is manifest. For instance, I once paid a good round sum for a bush of blood-red *Rhododendron arboreum*; waited five or six years before it flowered, till at last it produced some pure white trusses!

Probably we shall never see in this country the Virginian fringe tree—*Chionanthus Virginica*—dressed out with its feathery panicles with a profusion that turns this round-headed little tree into the semblance of a dish of whipped cream. That is how it is said to behave in its native Eastern States and in Southern Europe; here, in western Scotland, although it grows vigorously, in ordinary seasons it bears just enough flower to make one wish for more. The botanical name of this genus—*Chionanthus*, meaning snow-white flower—is very apt to be confused with that of a plant belonging to a wholly different order, namely the winter sweet—*Chimonanthus fragrans*, meaning the fragrant winter flower. Fragrant it is beyond ordinary, and justifies its name by opening its curious fleshy petals during the shortest days. It is usually

Some Flowering Shrubs

seen trained on a wall, but is said to be quite hardy in the open at Kew. Another winter flower is the Californian manzanita—*Arctostaphylos manzanita*—one of the rarest of shrubs to be seen in British gardens, probably from the difficulty experienced in propagating it. It is well worth an effort to establish it, for it is strikingly beautiful when sheeted with its white heath-like bells,[1] but it is somewhat miffy when young.

When the Mexican orange-flower—*Choysia ternata*—was first brought to this country it was treated as its origin seemed to indicate to be necessary, as a conservatory or cool house shrub; but it has proved quite hardy in all but the coldest and most exposed districts, a notable addition to our evergreens, with leaves scented like eucalyptus and flowers smelling of orange blossom.

M. Lemoine of Nancy has elaborated so many lovely hybrids among the species of *Philadelphus* (popularly and perversely spoken of as *Syringa*, which is the proper family name of the lilacs) that the best way to make a choice among them is to visit a good nursery when they are in flower. Special mention, however, must be made of one species not so often seen as the others, namely *P. microphyllus*, a shrub usually growing three or four feet high, flowering in July and diffusing a pine-apple fragrance to a distance of several yards.

---

[1] Mr. Bean describes the flowers as deep pink, but those of the largest plant I have seen, in the Edinburgh Botanic Garden, are pure white as also are those of its offspring here.

Flowers

The showiest of the American brambles or raspberries is *Rubus deliciosus*, but it has not behaved nicely with us, most of our plants having disappeared. *R. Nutkanus*, on the other hand, spreads pretty freely in competition with our native woodland undergrowth. It is hardly a fitting subject for gardens, though its leaves turn to fine colours in autumn. We have lost *R. spectabilis* with its reddish-purple flowers and excellent yellow fruit, not because of any difficulty on the part of this nice raspberry, but because it got crowded out and forgotten.

I shall lose my way presently among good things from North America, so let me bring the list to a close with a notice of three valuable Californian plants. *Garrya elliptica* is an old favourite which cheers us through the depth of winter with its quiet grace. It is usually treated as a wall plant, but I have in mind a vigorous bush full ten feet high growing in the open as far north as Fochabers in Elginshire. One should be careful to plant only the male of this shrub, which bears much longer catkins than the female. These catkins bear a singular resemblance, even in detail, to the slender garlands which form such a frequent feature in the plaster designs used by the Adam brothers for the decoration of cornices, chimney-pieces, etc., and it has been suggested that these famous architects took *Garrya* as a model; but that is not possible, forasmuch as this shrub was first brought to England by Douglas in 1828, and none of the Adam brothers survived the eighteenth century. Of *Fremontia Californica* I cannot speak from experience of its behaviour

Some Flowering Shrubs

in Scotland, but having seen it loaded with golden blossom on a wall at Glasnevin and in a windy corner at Leonardslee, I cannot doubt that it would thrive as vigorously on the west coast as its compatriot *Carpenteria Californica*. Our only reason for not having it is want of wall space.[1] *Carpenteria* is a good wall plant as far north as Western Ross-shire; its great white flowers look their best here when *Tropæolum speciosum* hangs its crimson garlands among them. The said *Tropæolum*, which our cottagers speak of as "petroleum," runs about everywhere, even to the top of a twenty-five foot holly. It comes from that floral treasure-house, Southern Chile, and finds our cool soil and ample rainfall so much to its liking that it is only its exceeding grace and beauty that saves it from being placed in the category of troublesome weeds, for it creeps its way into all our borders and kindles its flames upon all sorts of plants. It generally behaves as a herbaceous plant, dying down in winter; but when it gets among the branches of an evergreen, as in those of the holly aforesaid, it endures through the winter and starts flowering again in June. It has escaped from the confinement of park and garden into a wayside hedge, giving us a hint which we have taken by planting roots of it beside woodland walks.

Mention of this *Tropæolum* leads us to South America, whence a wealth of fine plants have been brought to our shores, especially from Southern Chile, a region which one longs to have fully explored, as no

---

[1] Since writing this I have seen a fine plant of *Fremontia* in full flower in Mr. M'Douall's remarkable collection at Logan, Wigtownshire.

Flowers

doubt it will be some day, so many things already introduced having prospered so heartily under British conditions of soil and climate, especially in the westerly parts of our island. For instance, among the multitude of species of barberry discovered of late years in Asia, there is none comparable in brilliance with *Berberis Darwini*, which Charles Darwin collected in 1835 during his voyage as naturalist in H.M.S. *Beagle*. Given reasonable shelter from wind, this splendid evergreen may be accounted hardy even in the colder parts of the realm, and in the milder districts rises to a height of 12 to 15 feet. By a happy accident it became crossed in the nursery grounds of Messrs. Fisher and Holmes at Sheffield, with the low-growing *B. empetrifolia*, also a Chilean shrub, whereof the result took the form of *B. stenophylla*, pronounced by Mr. W. J. Bean, Curator of Kew Gardens, to be " undoubtedly the most useful and beautiful of all the barberries." It is difficult to decide in which stage of flowering *B. Darwini* is more pleasing to the eye—before the blossoms open, when it is thickly hung with ruddy tassels of buds, or after the whole bush is ablaze with flaming orange flower. There is a third species of barberry from Chile—*B. buxifolia*—which, though less dazzling than *B. Darwini*, is, in my poor judgment, of more engaging beauty. It grows to an even greater size; there is a bush at Monreith measuring 126 feet in circumference and 18 feet high, presenting a remarkable appearance when hung in April with millions of little golden globes, filling the air with honeyed fragrance a long way to leeward. It is one of three shrubs

Some Flowering Shrubs

which, more than any others, perfume the breath of spring. *Azara microphylla* is another, also from Chile, a pretty evergreen which has already developed into a small tree. Its myriad tiny golden flowers are modestly born on the underside of the branchlets, and diffuse a spicy scent that has earned for the plant from children the name of " the chocolate tree," the odour being that of vanilla, with which chocolate is commonly flavoured. The third chief source of vernal incense is *Erica arborea*.

The delicate mechanism provided to secure cross-fertilisation in these and all other species of barberry, including *B. vulgaris* which is doubtfully claimed as indigenous to Britain, is so well known to every student of botany that I hesitate before drawing attention to it, and would not do so in this place had I not found that some of my fellow-amateurs have not noticed it. The flower consists of six sepals and six petals, all highly coloured, the petals of some species being of a slightly richer hue than the sepals. At the base of each petal are two nectaries, and in each petal reclines a stamen. In the centre rises the stout green pistil with a flat disc-shaped top, the circular edge of which is the stigma. Nothing happens till a bee or other insect alights on the flower and touches the stamens which, being highly sensitive, spring up smartly and discharge their pollen on the intruder, who will leave some of it on the stigma of the next flower visited.[1] One may witness this interesting action of

---

[1] The corresponding movement of the stamens in *Salvia* and *Roscoea* is mechanical, through the action of levers; but in the barberries the bases of the stamens are truly sensitive to touch.

Flowers

the stamens by touching them lightly with a pin or a stem of grass. In compliance with the rule of priority Sweet's name *Berberis dulcis* has been suppressed in favour of Lamarck's *B. buxifolia*, which is not nearly so appropriate, as the leaves, each with its needle point, are not very like those of box, while the epithet *dulcis* is well merited alike by the fragrance of the flower and the sweetness of the fruit. It is the only barberry, so far as I know, whereof the berries are agreeable to the palate without cooking; but this is so well known to blackbirds and thrushes that one seldom gets a chance of sampling them.

This brings to mind a pretty display which I witnessed one spring evening from the library window. A brace of cock golden pheasants—*Chrysolophus pictus*[1]—were busily engaged picking the yellow blooms off a bush of mahonia. So brilliant was the group in the rays of a westering sun that I could not find it in my heart to disturb them. Golden pheasants, which are never commonly confined to an aviary in this country, are thoroughly naturalised in our woods, and are welcome visitors when they display their fantastic plumage in the flower garden. Not so the common pheasant. A bull in a china shop—a pike in a trout stream—a schoolboy in the still-room—may be reckoned types of destructive energy; but for devastating plunder of

---

[1] This is another instance of the sacrifice entailed under the rule of priority in nomenclature. *Chrysolophus pictus*, meaning the painted goldcrest, is certainly descriptive as far as it goes, but it falls far short in poetic synthesis of the name which it supersedes—*Thaumalea picta*, the painted wonder-bird.

Some Flowering Shrubs

precious things a cock pheasant in a spring flower-border is hard to beat. Nor is this most uxorious fowl content to breakfast alone at this season. He must needs bring his dusky squaws to share his repast on crocus, dogtooth violet, fritillaries, *Saxifraga burseriana* and such toothsome titbits. Wherefore, so often as a cock pheasant flaunts his finery in the flower-garden, there ensues a stealthy stalk, a deed of blood and precipitate flight of his feathered harem.

One does not see the North American *Berberis pinnata* (*Mahonia fascicularis* of de Candolle) grown so often as it deserves to be; though perhaps it may escape notice in a young state from its resemblance to the common mahonia.—*B. aquifolia*. It is, however, a much more stately plant than that well-known shrub, for it rises to a height of twelve or fourteen feet with a stem girth of eighteen inches. It should be given a fairly sheltered position to protect its fine foliage from biting winds in March, when its branches become crowded with fragrant yellow blossom. The largest specimen of this shrub that I have seen is at Glasserton, in Wigtownshire. Six years ago I measured its circumference at 135 feet, notwithstanding that one side has been severely cut back to clear a gravel path and another side abuts upon a yew hedge. I could not get the height of this mighty bush accurately, but it cannot be short of eighteen feet.

The European myrtle—*Myrtus communis*—may no doubt be grown in the open in such exceptional places as the Scilly Isles and Channel Islands, though

Flowers

I do not remember to have seen it there. Even in other mild districts it requires the shelter of a wall; but its Chilean counterpart—*M. luma*—flourishes in the open in many parts of the country, forming a dense evergreen bush up to fifteen feet high and covering itself with white flowers in July and August, these being followed by a crop of agreeably-flavoured black berries, which would probably make very good jam. Counterpart, have I said? Yes, in all respects but one, and that must be reckoned deplorable: the flowers of the Chilean myrtle are almost scentless and its leaves are devoid of the aromatic virtue of its European relative. In another respect the western plant has the advantage. Its bark is very smooth, of a bright cinnamon hue, so that, if you have several plants, it is well to cut away some of the lower branches from one or two in order to reveal this feature.

It is high time that a simple English name were devised for that splendid Chilean evergreen which used to be called *Crinodendron Hookeri*, Gay, a title now altered to *Tricuspidaria lanceolata*, Miguel—pretty mouthfuls both of them! The allusion to flower-structure in *Tricuspidaria* is not very obvious, but *Crinodendron* means "Lily tree," which might conveniently serve us in everyday use, forasmuch as no extravagant measure of poetic license is required to recognise it as fairly descriptive of a specimen standing within fifty yards of my writing table, fifteen feet high, thickly hung on this June morning with pendent urn-shaped blossoms of a rich crimson hue. It is

Some Flowering Shrubs

perfectly hardy in mild districts near the sea, and makes a fine woodland ornament; but it will not suffer exposure to cutting winds, for the flower-buds are put forth on long footstalks nine or ten months before the flowering season. In colder parts of our country it may be grown, not trained to a wall but planted under its shelter, which it well deserves, being one of the very finest evergreens ever brought to this country. It ripens plenty of seed in the shape of pure white hard globes as big as a marrowfat pea, enclosed in a leathery capsule. If sown as soon as the capsule gapes, the seeds germinate very quickly. The other species, *T. dependens*, with white flowers, is of inferior beauty and we dismissed it.

A noble compatriot of *Tricuspidaria* is *Eucryphia pinnatifolia* which covers itself with milk-white bloom after the other has gone out of flower (Frontispiece). Both plants enjoy the same conditions of shelter, moisture and a peaty soil, and are meet companions, for although this species of *Eucryphia* is deciduous, in summer garb its grass-green foliage contrasts admirably with the sombre evergreen of *Tricuspidaria*. But see that you set them not too near each other or you will be landed in the same dilemma that confronts me every time I look forth from the library window. A plant of each were set when eighteen inches high with a ten-feet void between them : they are now both fifteen feet high ; the void has disappeared and they are interfering with each other's symmetry. One or other will have to be sacrificed so soon as I can muster hardihood for the deed. The other Chilean species, *Eucryphia*

Flowers

cordifolia, is reported to be less hardy than the aforesaid, but it has never suffered here. It is evergreen and opens its lovely white flowers somewhat later than *E. pinnatifolia,* continuing to produce them until the end of October. The seed takes nearly two years to ripen, but the plant sometimes produces suckers and may be grown also, but uncertainly, from cuttings. A third species, *E. Billarderi,* from Tasmania, has not yet flowered at Monreith, but a plant given to me by Sir John Ross of Bladensburg is making good growth, and I am told to expect it to produce a very pretty bloom. The significance of the name *Eucryphia,* meaning " well-covered," is not apparent at first sight. It has been conferred on the genus because of the little brown caps that cover the tips of the flower-buds of *E. pinnatifolia* and are thrown off as the flower opens.

In districts where it can suffer the climate without flinching, there is no Chilean shrub that grows more rapidly and makes a more lavish display of blossom than *Abutilon vitifolium,* rising to a height of twenty feet or more and spreading its skirts equally wide. The typical colour of the large blossoms is rosy lavender or palest violet, but there is a nice white-flowered variety ; also one of a disagreeable reddish tint, which should be rigidly suppressed. This shrub ripens plenty of seed, whereof it is advisable to sow some occasionally, because its term of life is very uncertain. It may live in perfect vigour for ten or twenty years before suddenly expiring, perhaps through exhaustion from the strain of producing so

Some Flowering Shrubs

much flower and seed. On the other hand, *Abutilon Megapotamicum*, also from South America, has never ripened seed for us, not from any parsimony in the matter of flowers, for of all the shrubs grown here it is the most lavish of bloom, beginning—it is hard to say when it begins, for if the winter is mild there is scarcely any time of the year when it is without some of its quaint blooms, suspended like rows of little two-inch dolls, each with a bright yellow petticoat showing under a crimson cloak, the style and anthers serving as neat purple footgear. It is not so hardy as the other species, but in mild districts lives quite happily against a wall in any aspect and is very easily propagated from cuttings. This is another instance of the priority rule displacing a good descriptive specific name in favour of an earlier one of feebler significance. Saint-Hilaire christened the species *A. vexillarium*—the standard-bearer—most appropriate for a shrub that flaunts the Spanish national colours—scarlet and yellow—in South America; but he had been anticipated by an earlier botanist who, having found it growing on the Rio Grande, named it *Megapotamicum*, and we have to follow suit.

It would entail a lengthy chapter to notice even half the beautiful things which have come to us from Chile, and I shall only claim space to mention one more, to wit, *Desfontainea spinosa* (Plate VI). Here again we have a plant reminding one of the Northumberland miner who went into a draper's shop in Newcastle to buy a neckerchief. The assistant behind the counter produced some for approval, chiefly blue,

Flowers

purple and green. " Na, na," exclaimed the customer; " I want nane o' your gaedy collors. Just gie me plain reid and yalla." Plain red and yellow are the tubular flowers of *Desfontainea*, highly varnished to enhance their brilliancy. In shape and colour the leaves are so like those of the holly, that visitor's often express surprise at beholding a holly so strangely decked. Nearer inspection would show them that whereas the leaves of holly are alternate, those of *Desfontainea* are opposite. *Desfontainea* seems to thrive more vigorously in the west of Scotland than in any other part of the— it was on the tip of my pen to write United Kingdom— of the British Isles; but, strange to say, it cannot be got to grow in Cornwall. So I am told by the experienced owner of a magnificent collection of plants in that happy land, who has planted it repeatedly, but has never been able to keep it. Here it ripens plenty of seed, but it is a rare thing for a self-sown seedling to survive the necessary weeding of garden borders or the uprush of vegetation in the woods. There is a very large specimen at Stonefield on Loch Fyne, round which I paced fifteen years ago and found it to be 75 feet in circumference, though it had been sorely shorn back on one side to clear a gravel path. Marry! but I would have had that path shifted or obliterated rather than suffer mutilation of that splendid plant, which, when it flares up in a blaze of scarlet and yellow in July, must be no feeble simulacrum of the bush out of which the angel spoke to Moses. I guessed the height of it to be fourteen feet. There is a still larger specimen at Corswall near Stranraer.

PLATE VII

⅔ natural size   DESFONTAINEA SPINOSA   20 August, 1920

Some Flowering Shrubs

The colours—scarlet and orange—which prevail so frequently in the flowers of Chile do not seem to have received the attention from botanists that they invite. The climate of Southern Chile is described by travellers as being very similar to that of our west coast districts—mild winters with much rainfall and cool summers. In such measure, therefore, as temperature, sunlight and moisture may be accounted as influencing the colour of flowers, it might be expected that these agents would effect some similarity in the hues of British and Chilean plants. That they have not done so appears in the fact that there is no wild flower with an orange corolla indigenous to the British Isles, and none with a scarlet corolla except the corn poppy and the shepherd's weather-glass (*Anagallis arvensis*), both of which are weeds of cultivation, never appearing except in tilled land, and having, as George Bentham remarked, " accompanied man in his migrations over a great part of the globe." It is impossible now to ascertain the place where these two plants originated; but it is most probable that they were brought into the British Isles from the Continent.[1] All our native red wild flowers have an admixture of blue in the tint, producing in such species as *Orchis mascula* and *Geranium sanguineum* unmitigated

---

[1] In the flowers of bird's-foot trefoil (*Lotus corniculatus*) the standard is sometimes splashed with vermilion on the outside, but this is far from constant. There is a garden variety of the Welsh poppy—*Meconopsis cambrica*—with good orange flowers. I have never happened to meet with *Adonis vernalis*, which has scarlet petals and is included in the British flora; but I understand it to be a rather infrequent weed of cultivation, probably introduced.

Flowers

magenta. In the Chilean flora the tendency is all the other way. The genus *Buddleia*, which in Asia displays purple, pink or white flowers, is represented in Chile by the orange-flowered *B. globosa*. In the old world *Lobelia* tends to purple and blue, while Chile produces the stately *L. tupa*, though it must be admitted that its scarlet is eclipsed by that of the North American species, *L. fulgens, splendens* and *cardinalis*. Chilean brilliancy culminates in the scarlet sprays of *Embothrium coccineum*, and the vehemently vivid orange vermilion of *Hippeastrum (Habranthus) pratense*.

Of the *Escallonia* family *E. Philippiana* is the hardiest of the deciduous species and *E. rubra* of the evergreen, but the latter is far inferior in beauty to *E. macrantha*, which flourishes anywhere near the sea. The tallest of the genus is *E. Exoniensis*, a fine evergreen growing twenty feet high, a hybrid between *E. pteroclados* and *rubra*, owing the roseate tinge in its white flowers to the latter parent. The best of the genus is another hybrid—*E. Langleyensis*—offspring of *E. Philippiana* and *punctata*. Rising to ten or twelve feet in height and spreading very wide it makes a splendid appearance after midsummer when its long arching sprays are beaded throughout their whole length with bright crimson flowers. But to secure the full effect plenty of room must be accorded to this shrub, which never looks better than when isolated on a lawn. It is with some misgiving that I admit *Calceolaria violacea* to a list of hardy plants, for hardy it cannot be reckoned except where winters are moderate; but it is such a delightful shrub, so

Some Flowering Shrubs

generous with its quaint little cup-shaped flowers elaborately painted inside with gold and purple that it deserves a place at the foot of a wall where, even if it is cut down by frost occasionally, it will spring again and flourish in a milder season. It lasts fully six weeks in bloom.

The shrubs of New Zealand are remarkable for the large proportion of evergreen Composites among them. Of the daisy bushes (*Olearia*) the only one that can be pronounced generally hardy in Great Britain is *O. haasti*, from the South Island, which hides every leaf with its white flowers in August; and even that was cut to the ground at Kew in the exceptional cold of 1894-5. But in maritime districts many other species ought to be grown for their beauty. *Olearia nitida* loads itself with white corymbs of flower in May and grows to a height from eight to ten feet. *O. macrodonta* follows in June, attaining the dimensions of a small tree, and so does *O. ilicifolia*, which prolongs the display through July. These three species are indigenous to both the North and South Islands of New Zealand, and in getting plants or seed it is well to ensure having them from the South Island which, being further from the Tropic of Capricorn, enjoys a more temperate climate. *O. Traversi* from the Chatham Islands is the tallest of the genus, distinguished as the only one of the large-leaved species, except the little-known *O. Buchanani*, that has opposite leaves. The foliage is silvery-grey and very attractive, but the flowers are a fraud. When planted here it shot up to a height of eight feet in two years, and every branch

Flowers

was thickly hung with panicles of flower buds. My stars! methought, here is something worth waiting for; but when the buds opened in June, lo! they were rayless and the discs were dingy yellow, no whit more ornamental than our common groundsel. *O. Traversi* would probably make a serviceable hedge, but so would many a better thing, so on the principle of growing only the best best, it has been added to our gardening *index expurgatorius*. *O. semidentata*, also native of the Chatham Islands, is very distinct, with linear lanceolate leaves densely clothed underneath with white wool, $1\frac{1}{2}$ to $2\frac{1}{2}$ inches long, and with handsome flowers borne solitary 1 to $1\frac{1}{4}$ inch across the rays, which are pale lavender set round a violet disc. The flowers have the peculiarity of increasing considerably in size after opening. Mr. Cheeseman puts the height of this bush at one to three feet, but here it has already risen to between four and five feet, and is still going strong. *O. Chathamica* resembles *O. semidentata* pretty closely, but the leaves are longer and broader and the flowers are larger. About its merit compared with the other I am not qualified to speak, seeing that the only specimen we had here "died on me," not from winter cold, but in midsummer. So did the largest-flowered species of the genus, *O. insignis*, a most desirable dwarf shrub to be tucked into a warm nook under a wall; but our winter wet was too much for it. There are fine specimens of this plant in Sussex gardens at a considerable distance inland. *O. stellulata* (still listed by some nurserymen by the older name of *Eurybia Gunniana*) is an Australian species, reputed to

Some Flowering Shrubs

be less hardy than some others; but it has endured 17° Fahr. of frost here without forfeiting its annual display of white flowers, which are larger than those of most other daisy-bushes.

The last of this family that I shall name is *O. nummularifolia*, though why Sir Joseph Hooker devised a hepto-syllabic name for an inoffensive shrub the present deponent doth not know. *Nummularius* is Latin for a banker or money-changer, so that does not explain his meaning; and if an analogy is indicated with the leaves of creeping Jenny—*Lysimachia nummularia*—it is somewhat far-fetched, for the leaves of the shrub bear but slight resemblance to those of the herb, are of much firmer texture and are set in quite a different manner on the stem. Well, this shrub, to denote which we, having coined no English name for it, must employ twelve syllables, has earned our favour by reason of its close olive-green foliage, distinct in habit from all other evergreens, a cheerful object in winter.[1] But the brightest of all green things for mild districts, one that, like all the *Olearia* family, enjoys the buffeting of sea winds, is the New Zealand " kapuka "—*Griselinia littoralis*—one of the Dogwood order, invaluable as a shelter plant for more sensitive things. It is not hardy at Kew, which is surprising, for here it came unhurt through the long frost of 1895, the only winter in my life-time when the mercury fell here below zero. In that season, however, we had a deep

[1] All the *Olearias* ripen and discharge clouds of seed, which float away on their little parachutes; but *O. nummularifolia* is the only species of which I have found self-sown seedlings in the borders. Outside the garden, of course, the undergrowth is too lush to allow such small seed to germinate.

Flowers

cover of snow, very unusual on the west coast, which protected many things that must otherwise have succumbed to cold. This *Griselinia* is by nature a tree up to fifty feet high; but through being propagated from cuttings in this country it is usually seen as a large bush, just as the common cherry-laurel—*Prunus lauro-cerasus*—has acquired a shrub-like habit owing to similar treatment.

In this country the showiest native *Senecio* we can boast is the too common ragwort—*S. Jacobœa*—but there are several very handsome shrubs of that genus in New Zealand, hardly to be recognised by casual visitors as members of the same clan as our common groundsel, which may be grown successfully in many parts of Great Britain and Ireland. *Senecio Hunti*, for example, is reported as growing twenty feet high in its native Chatham Islands, and here is already one-fourth of that height and the same in diameter from cuttings given to me by Sir John Ross of Bladensburg three or four years ago. It is a fine sight when covered at midsummer with large panicles of yellow flowers, resembling those of *S. Greyi*, a lowlier and more spreading shrub. To the same generous donor I owe the white flowering *S. Hectori*, with leaves six to twelve inches long and three or four inches broad, a handsome bush. *S. compactus* resembles *S. Greyi* on a smaller scale, but is hardly worth growing if one has the other. The finest of this group is said to be *S. Kirkii*, which grows well with Mr. Boscawen at Ludgvan, near Penzance, but we have not tried it here, and, coming as it does from the

Some Flowering Shrubs

North Island, is likely to be more tender than *S. Hectori.*

About the Veronicas of New Zealand my words must be few and uncertain, for they seem as perplexing as the willows of the northern hemisphere. Cheeseman enumerates eighty-four species, many of them breaking into several varieties, besides natural and manufactured hybrids. *Veronica Traversi*, for instance, grows here nine feet high and very wide, whereas Cheeseman describes it as a small bush, two to five feet in diameter. In another work, however, he says of *V. Traversi* "if I am correct in my identification of that very problematical plant."[1] Then I am directed now to regard as *V. salicifolia* what we have long looked upon as *V. parviflora*. If these are one and the same species, they are indeed very distinct varieties, for *V. salicifolia*, the handsomer of the two, bears long pendant racemes of flower and finishes flowering by the beginning of August, while the other shrub holds its shorter racemes almost horizontally and continues to flower till stopped by frost in late autumn.

Sophora tetraptera in all its varieties is a beautiful shrub or small tree, demanding the protection of a wall, I should think, even in the milder parts of England and Scotland. Here it produces its large flowers, old gold in colour, very freely, and ripens seed enclosed in curiously-shaped pods. The variety *microphylla* is said to be hardier than the type, but the flowers are somewhat smaller.

[1] *Illustrations of the New Zealand Flora*, vol. ii. 149.

Flowers

Of the genus *Metrosideros* I cannot speak from personal experience of its behaviour in Western Scotland; but the splendid display with which some of its species illumine gardens in Cornwall and the Scilly Isles make me ashamed of not having had the sense to adventure with them here long ago.

"Of the eleven species of *Metrosideros* growing in New Zealand," says the Rev. A. T. Boscawen, "only two or three, so far as I have experienced, are hardy in Cornwall. The best of these are *M. robusta* and *lucida*. Considering their beauty I wonder they are not more often grown. Of all New Zealand shrubs, I know of none more desirable."[1]

Cheeseman describes *M. diffusa* as the most brilliant species; and *M. tomentosa* flourishes gloriously at Tresco.

The Australian bottle-brushes are often confused with *Metrosideros*, but they belong to a different genus—*Callistemon*, containing some beautiful shrubs with cymes of crimson flowers. Probably some of the species are hardier than is generally known. *C. speciosus* thrives here in the open border without any protection. Beginning to flower about midsummer, it continues as a brilliant object until dimmed in the dark days of October. It is one of the very choicest things for planting in the milder parts of our country; but those who observe the salutary practice of snipping off the faded trusses from rhododendrons, thereby relieving the plants from the exhausting function of seed-bearing, must by no means apply similar treatment to *Callistemon*, because of a peculiarity in the

[1] *Journal of the Royal Horticultural Society.*

PLATE VIII

½ natural size   'KOWHAI'   31 May, 1922
CLIANTHUS PUNICEUS

Some Flowering Shrubs

growth of that species. The characteristic "bottle-brush" consists of a dense cylinder of flowers encircling the twig. As these fade the seed-vessels swell, forming a persistent woody column, from the summit of which projects a pointed bud to form a new shoot in the following season. If, therefore, the old flowers are removed, the plant is prevented from forming its fresh growth. *Clianthus puniceus* (Plate VIII) comes from the North Island of New Zealand, nevertheless it grows well in many parts of Great Britain with the protection of a wall, producing its glorious scarlet lobster-claws in such profusion as few other plants can achieve. In richness of flower and foliage it has few equals, and though it may be cut to the ground in a winter severe beyond normal, it will spring again and recover lost height very fast. Moreover, it is very easily propagated from seed or cuttings, which it is well to have in stock, for, like some other soft-wooded plants that flower extravagantly, it is apt to die suddenly from exhaustion. Mr. Cheeseman says that it has become well-nigh extinct as a wild plant, but it is commonly cultivated as a garden plant in New Zealand. There is a dingy white variety that is not worth attention.

Of *Gaya* (*Plagianthus*) *Lyalli* there is more than one variety; the showiest is said to be the one distinguished as *ribifolia*, whereof the leaves, being smaller than those of the type, allow the flowers to be better seen. We have here only young plants of that variety which have not yet flowered; but the broad-leaved kind is a beautiful tree, and, being late in

Flowers

coming into leaf, has nothing to fear from spring frost. We have grown at Monreith for many years both *Notospartium Carmichaeliæ* and *Carmichaelia australis*, the pink and purple brooms of New Zealand, but although they have endured 17° Fahr. of frost without flinching, neither of them has yet given us a flower, so I have not been able to verify the points which distinguish these two very similar genera from each other.

The New Zealand manuka—*Leptospermum scoparium*—is a charming shrub for mild districts, wreathing its slender sprays with crowds of white flowers in June; while the variety *Nicholli* bears bright crimson ones and *Chapmanii pink* ones. We grow two other species—*L. ericoides* and *lanigerum*—the family likeness running strong among them; all being equally impatient of overhead shade and enjoying open, windy positions. Where the climate is too severe for them in the open they are well worthy of the shelter of a wall.

Coming now to Asiatic plants, every country-bred child has learnt to love the laurustinus (*Viburnum tinus*) from Southern Europe for the quiet gaiety of its blossom in the dark days; but everybody does not know the variety named *lucidum* because of its shining leaves. It bears much showier flowers, and, although not quite so hardy as the type, it may be planted with confidence in most places near the sea. The laurustinus is evergreen; but the queen of the family—*V. Carlesii*—unrobes in winter, donning her verdure afresh in April as a setting for clusters of deliciously-

Some Flowering Shrubs

scented flowers, rosy in the bud, chalk white when open. This treasure only landed in this country in 1902, and I really think that if I were limited to the choice of a single shrub from among all the fine things introduced for the first time from China during the present century (*not* including rhododendrons), this would be the one. It has proved quite hardy, and only in one respect have we found special attention required in its cultivation. Nurserymen often supply it grafted on the common wayfaring tree—*V. lantana* which, being a more vigorous shrub than the newcomer, persists in sending up strong suckers. As the leaves of both species rather closely resemble each other, the robbers are not easily detected, wherefore it requires much vigilance to save the precious plant from being evicted by them.[1] Similar care is necessary for the safety of *V. tomentosum* and its truly delightful varieties *Mariesii* and *plicatum*, which are well worthy of inclusion in the most limited collection of shrubs. *V. rhytidophyllum*, introduced by Wilson in 1900, is certainly an interesting evergreen, but, in my poor judgment, of indifferent merit as a decorative plant. The flowers are freely produced in umbels like those of the common elder; the foliage may be described as handsome, and it grows rapidly to a height which I am unable to specify, because we had to uproot the only two specimens we had here before they exceeded eight feet to save better things from being crushed

[1] Thus far had I written before receiving from a friend a gift of the still more recent importation from China of *Viburnum fragrans*, which, if it be not the peer of *V. Carlesi*, is at least a worthy second. It is reported to be the hardier of the two.

Flowers

out of existence. *V. utile* and *crassifolium* we threw out also, as unworthy members of a noble family.

The pearl bush—*Exochorda grandiflora*—deserves more attention than it has received from garden owners. True, it is no novelty, having been one of Fortune's finds in China about seventy years ago; but one does not often meet with it. It grows ten feet high or more, bears showy racemes of pure white flowers in May and ripens plenty of seed. *S. Giraldi* is reported by those who have seen it (as I have not) to be even handsomer. It was first brought from China in 1909 and should be worth enquiring for. *Xanthoceras sorbifolia* is from Northern China. Our best plant at Monreith is placed at a disadvantage through being too near a fine horse-chestnut, a tree to which it is closely akin and with which it flowers simultaneously. It must be owned that the chestnut far excels it alike in stature, beauty and duration of blossom. Our plant is only some twelve feet high and would be more admired when in flower if it were not so placed as to look like a poor relation of the other. We prize it, however, for one may visit a hundred gardens and not see another specimen; indeed, the only place where I happen to have noticed it is the Botanic Garden at Glasnevin. The flowers stand in erect panicles, with a rufous or crimson stain at the base of each petal. They are followed by capsules as large as those of the horse-chestnut.

When Sir Joseph Hooker pronounced *Buddleia Colvillei* to be the handsomest of all Himalayan flowering shrubs, he probably meant to exclude rhododen-

PLATE IX

⅜ natural size   BUDDLEIA COLVILLEI   14 July, 1919

Some Flowering Shrubs

drons; otherwise it is difficult to conceive his preferring it to such a grand thing as *R. Aucklandi* (I cannot train myself to speak or think of this species under the name now prescribed for it—*R. Griffithianum*), for when that is in good condition and full bloom this peerless shrub fairly earns the encomium passed by Horace upon the young Marcellus:—

> ... Micat inter omnes
> Julium sidus, velut inter ignes
> Luna minores.[1]

Probably this queen of the race of *Buddleia* flowers more profusely among its native mountains than it ever does in this country; then, indeed, it must be a magnificent sight, for it attains the height of forty feet, and the flowers, much larger than those of any other of this genus, are rose-coloured,[2] of a waxy texture, and are borne in pendent panicles six or eight inches long. Growing against, but not trained to, the wall of a laundry, we have it here twenty feet high. It flowers pretty freely in most seasons, but not so profusely as to prevent us wishing for more, and it is a misfortune that its blossoms fade very quickly when cut and placed in water, as I found to my inconvenience when painting the original of Plate IX. Everybody now grows *Buddleia variabilis* in one or other of its fine varieties. Well that it is so popular, were it only for the irresistible attraction it presents to Red

---

[1] And like the moon the feebler fires among
Conspicuous shines the Julian star.
(Francis' translation.)

[2] There is a claret-coloured variety, far inferior, methinks, to the usual pink.

Flowers

Admiral and Painted Lady butterflies. Professor W. W. Smith tells me that *B. alternifolia* is a more beautiful shrub than *B. variabilis*, its long willowy shoots being richly wreathed with violet flowers. He showed me the plant in the Edinburgh Botanic Garden—about eight feet high; but it was not in flower at the time.

Fifty years ago, or thereby, I was prowling through the ample nursery-grounds of Messrs. Cunninghame and Fraser, in Edinburgh, when my attention was drawn to some bright golden blossoms in a neglected corner among some gooseberry bushes. Closer inspection revealed a bush of a kind new to me—*Hypericum Hookerianum*, a Sikkimese subject.[1] I rescued it from its forlorn situation, and have never been without plenty of it ever since, for it is a very choice St. John's-wort, growing four or five feet high, and bearing its rich yellow flowers throughout August and September. Nearly allied to it is the Japanese *H. patulum*, which continues longer in flower than the other, persisting till autumn frost puts an end to the show. Dr. Augustine Henry discovered in China a variety now known as *H. patulum Henryi*, much handsomer than the Japanese form, with flowers of deeper gold, but its flowering season is not so prolonged. As to hardihood, *H. Hookerianum* may safely be planted almost anywhere; *H. patulum* has nothing to fear in the west; about Dr. Henry's variety I can only speak confidently of its vigour in our own garden; but it is probably capable of enduring more cold than the type.

Clerodendron trichotomum is a handsome shrub or

---

[1] Also known by the equally cumbrous synonym—*H. oblongifolium*.

Some Flowering Shrubs

small tree where it gets enough sun; there used to be a splendid specimen in Messrs. Veitch's nursery, now no more, at Coombe Wood, besides others at Abbotsbury and elsewhere; but although it is quite hardy, it comes into flower too late to be of much use in northern gardens. *C. Fargesi*, introduced from China in 1898, resembles it very closely, except that the calyx, instead of being red as in *C. trichotomum*, is green, which makes the bloom decidedly inferior to that of the other, and renders the plant unworthy of cultivation. *C. fœtidum* is a far finer thing than either of these, but being not so hardy, requires the protection of a wall. It bears beautiful corymbs of carmine flowers at the ends of the branches and its foliage is very handsome. In all three species there is the same curious contrast between the delicious fragrance of the flowers and the disagreeable smell of the leaves.

Most of the Dogwood family hail from North America; but *Cornus Kousa*, of Chinese and Japanese origin, is a small tree of much merit and, on the whole, is the best of the genus for British gardens, being the Asiatic counterpart of the American *C. Florida*, which we need not attempt, for though it will live in Britain, it never produces the lavish bloom that adorns the woodland of the Eastern States in May. The Himalayan *C. capitata* (usually known as *Benthamia fragifera*) makes a fine show in the southern and western counties, but frost in April or May sometimes destroys the promise of flower of *C. Kousa*. We used to prize the cornelian cherry—*C. mas*—very highly for the mist of tiny yellow flowers which it throws over its leafless

Flowers

branches in February; but that has been totally eclipsed in favour by the witch hazels, first by *Hamamelis arborea*[1] and later by *H. mollis*, which light up the darkest days of the year with a profusion of golden bloom. *H. mollis*, indeed, has proved itself one of the very choicest deciduous shrubs discovered in China during the last fifty years; it is absolutely hardy, grows at least ten feet high, and is generally in flower here before Christmas—always before the new year, *H. arborea* following about ten days later.

It is a sound maxim that bids a man speak only of what he knows; wherefore I shall say nothing about the fine genus *Forsythia*, except to express admiration for the beauty of *F. suspensa* and *intermedia* in the gardens of other people. We have never succeeded with it here, owing to carelessness, I suppose, combined with indolence, for it offers no difficulty in cultivation. Owing to the same vicious compound I am unable to speak from experience of the newer species of *Magnolia*; among the older ones there is none equal to *M. conspicua*, which flourishes here, and *M. parviflora*, which gives us some trouble owing to its twigs dying back.

The earliest of the Spiræa family is *S. arguta*, a hybrid between European and Asiatic species, which never fails to deck itself with blossom in April—pure white without the creamy tint which is so general among others of the genus. *S. bracteata* and *Van Houttei* follow in May, the latter being a hybrid of uncertain origin, and both being of the class that carry long

---

[1] An arborescent form of *H. Japonica*.

Some Flowering Shrubs

arching sprays richly set with white rosettes of flower. We had the two growing side by side, and the time came when one or other had to be sacrificed; it was no simple matter to decide between them, flowering simultaneously in June with white rosettes set on arching sprays; we hesitated for two summers, and ended by keeping *S. bracteata*, which was undoubtedly the handsomer. This bush, viewed at a distance, might easily be mistaken for a hawthorn in full bloom; but it lasts much longer in beauty than the may; indeed, it is one of the merits of this rosette class of Spiræa that the flowering period endures for three weeks or a month according to the weather. *S. canescens* and *S. Henryi* follow in July,[1] having the flowers similarly set in rosettes, but they are much taller plants growing to a height of twelve or fourteen feet, and requiring ample room for development. They are easily propagated from suckers. *S. brachybotris* I have not seen, but it has been described as a most desirable hybrid from *S. canescens* and *Douglasi*, the flower rosettes being bright pink instead of white. Very different in habit is *Spiræa arborea*, brought from China by Wilson and eclipsing in stature the older allied species *S. Aitchesoni* and *Lindleyana*. It is well named *arborea*, for it is practically a small tree, and bears magnificent panicles of cream-white flower in early autumn. Planted in cool moist soil with ample room to develop, it maintains a long succession of

---

[1] Mr. Bean describes *S. Henryi* as a June flowerer, as no doubt he has found this fine species to be at Kew; but in our northerly latitude it delays till July.

Flowers

bloom; but beware of placing it where it can overshadow less robust neighbours, for it may grow five-and-twenty feet high and arches over a very wide space. *S. Japonica* (syn. *callosa*) grows four or five feet high and produces flat corymbs of bright rose flowers from July till September; a very nice shrub for wood-land planting.

As there is nothing spiral in the general habit of any of the Meadowsweet family, it may have puzzled some amateurs in botany, as it used to do me, why the name *Spiræa* should have been applied to it. There was a time when the late Lord Avebury and I used to escape from the House of Commons (where he sat as Sir John Lubbock) for a ramble in the Kentish marshes or elsewhere. I never returned from one of those excursions without some fresh bit of knowledge, and it was on one such occasion that he showed me how the carpels of meadowsweet are twisted into a spiral; whence the name *Spiræa* for the whole genus.

In all the multitude of rose species that have been brought hither from the far East there is none to compare with *Rosa Moyesii*, one of the most remarkable acquisitions in recent times to our list of perfectly hardy shrubs. The intense velvety blood-red of its petals is unmatched in the flower of any other wild rose, and its delicately pinnate foliage adds to the charm of the plant. It is a rampant grower, throwing up shoots six feet long in a single season, so it must be given plenty of room. The variety named *Fargesi* is of equal merit, the petals being of a uniform rose-colour of peculiarly fine quality.

Some Flowering Shrubs

In *Osmanthus Delavayi* we have received another prize from China. It is not one of the "glut-and-famish" class that flower exorbitantly one year and sulk in the next, for this pretty shrub shrouds its dark evergreen leaves in every succeeding April with a veil of fragrant milk-white flowers. I know not to what stature it may attain; it inclines to a horizontal habit, and I have seen none over five feet high.

Pieris Japonica is not quite so hardy as the more commonly grown *P. floribunda*, but being in comparison to that rather dull species as a bottle of Hautbrion claret is to *vin ordinaire*, ought to be planted instead of it wherever the climate is fairly mild. *P. Japonica*, however, is in danger of being eclipsed by *P. Tawaniensis*, a Chinese newcomer, which, so far, has proved perfectly hardy here even in a very young state. *Deutzia longifolia*, introduced by Wilson in 1903, is one of the best of the genus; but the old *D. scabra* remains very hard to beat.

Piptanthus Nepalensis has been grown in England since 1821, requiring wall shelter in the colder districts; but here it flowers regularly in May in the open border, a pretty thing. Wilson sent home from China a closely-related species—*P. tomentosus*—which is said to be hardier than the other.

Several new species of *Indigofera* have been imported of late years from China; it may be hoped that some of them may be better suited to our climate than the pretty *I. Gerardiana*, which, if grown in the open, is generally cut to the ground in cold districts, springing again and bearing racemes of rosy mauve

Flowers

pea-flowers in August. Here we have it in bushes six feet high, and except the Himalayan whitebeam—*Pyrus vestita*—it is the latest of all flowering trees and shrubs to come into leaf. Neither of these are half-clad at this time of writing—28th June—but both of them are distinguished later by their beautiful, but very dissimilar, foliage.

Styrax Obassia and *S. Japonicum* are far too little known as beautiful flowering shrubs, quite hardy if given reasonable shelter from wind. *S. Hemsleyanum* is a more recent introduction, the best of the three species according to some, but I have not seen it in flower. They are all small trees and do best in what may be termed rhododendron soil, that is, light loam with sand and well-weathered peat; and, like rhododendrons, they do not thrive if it is too dry.

The list of desirable Asiatics might be prolonged indefinitely, but I will bring it to a close with a word of praise for *Fatsia Japonica*, pretty frequently known as *Aralia*. It is more often seen as a pot plant than in the open, but it is perfectly well able to endure a British winter, and its great, leathery, palmate, grass-green leaves stand buffeting by wind far better than might be expected from their appearance. In late autumn it produces handsome panicles of white flowers arranged in globular heads, the whole suggesting carved ivory. After a cold, wet summer like that of 1922, these flowers, in the north at least, do not appear in time to escape winter storms; but, given a good season, they are about the handsomest things in bloom in October and November.

Some Flowering Shrubs

With African shrubs I have little experience, so the less I say about them the better; indeed, there are but two that come to mind as fitted for our climate. *Genista virgata* is one, a native of Madeira which grows to a great size in our cloudy atmosphere. A bush of it twelve feet high and as much through makes a splendid show in July by covering itself with millions of small yellow flowers. The other is *Phygelius Capensis*, which is listed by the Kew authorities as a herbaceous plant, but which has all the qualities of a shrub, though in cold districts it is cut to the ground in each winter. Even in mild districts it is never seen to such advantage as when trained against a wall, a position whereof it takes such ready advantage as scarcely to be recognised as of the same species as the low, straggling bush which it makes in the open border. The best form of *Phygelius* has brilliant scarlet and yellow tubular flowers, and care should be taken to obtain it instead of a dull brick-coloured variety.

IV

Some Rhododendrons

AMONG all the natural orders of plants, there is none, not even the *Rosaceæ*, containing such a number and variety of beautiful flowering shrubs as the *Ericaceæ* or Heath family. One might go even further and affirm that the single genus *Rhododendron* exceeds any entire natural order in that respect, especially as, according to modern classification, Azaleas are now included in that genus. Nor are we yet in a position to gauge its full affluence. During the past thirty years the new species discovered in China by such enterprising collectors as Dr. Henry, Messrs. Wilson, Forrest, Farrer and others, and introduced to this country, amount to several hundreds, ranging from such humble creeping forms as *R. erastum* to the lofty *R. giganteum*, which is said to attain a stature of 80 feet. Many of these new species have already been tested in this country; it would be expecting too much that they should all prove worthy of places in our gardens and woodland, and undoubtedly several of them are of little more than botanical interest; but there are also many kinds of very great beauty.

Some Rhododendrons

The question is often asked whether any of them exceed in beauty some of the Indian and Chinese species with which cultivators in this country have been more or less familiar since the early years of the nineteenth century. It is too early to answer that question with any degree of confidence. The new species must attain mature size before they can be put in fair competition with the older ones; but this much I am prepared to affirm that it is extremely unlikely that any of them will excel or even equal the Indian *R. Griffithianum* (*Aucklandi*) in splendour of flower. Nevertheless, many of the new species from China are most valuable on account of their beauty, and of those which I have seen in good bloom the following seem to me indispensable for any amateur who aims at having a selection of the best.—*R. Fargesi, Soulei* (Plate X), *oreodoxa, oreotrephes, neriiflorum,*[1] *crassum bullatum, Augustini, calophytum, Yunnanense, argyrophyllum, hæmatodes, adenogynum, orbiculare, decorum* and *sinograndis*.

A considerable number of these newly-discovered species have already been proved to flower and flourish freely in the milder districts of the British Isles, including nearly the whole of our western sea-board from Cornwall to Ross-shire, subject to the indispensable condition that they can be given protection from wind. That is really a cardinal requirement for all Asiatic rhododendrons, except the small-leaved species of

---

[1] This is obviously a misnomer, and ought to be *neriifolium*. The flowers have no resemblance to those of an oleander, whereas the leaves of the two plants are somewhat like each other.

Flowers

humble stature, which in Western China take the place filled by heather in Europe, clothing the mountain-sides above the level of forest growth at from 10,000 to 15,000 feet. It will probably be found that the majority of Chinese species of rhododendron are capable of enduring all the frost they are likely to encounter even in the colder districts of Great Britain. Winter is not what these plants have most to fear in this country ; most of them come from high altitudes —5000 to 15,000 feet—where they are laid to complete rest for several months. The conditions are very different in an ordinary British winter consisting of long spells of mild, wet weather, causing the buds to swell and push prematurely. We are painfully familiar with the cruel frosts that so often come in March, April and even May, blackening and killing the young growth, whereby all prospect of bloom in the succeeding year is frustrated, for the subsidiary shoots that are put forth in summer carry no flower buds.[1]

Hence it follows that a mild winter is not favourable to Asiatic rhododendrons, unless it is followed by a genial spring without frost. The following experience shows how some of them actually prosper better in a cold district than a mild one. In the spring of 1921, *R. oreodoxa*, a lovely Chinese species, started growth in February in my own woods near the sea on the west coast. All its blossom was destroyed, together with all the young growth, by frost in the

[1] I have a note of 22° of frost on 20th May, 1894, the only season in which I have known the young growths on the ash to be destroyed. I had a fishing on the Lea near Hatfield in that spring, and all the shoots on the ash were turned as black as the buds had been whence they sprung.

Some Rhododendrons

first week of April. In consequence, the plants did not produce a single flower in the spring of 1922. On the other hand, at Dawyck in Peeblesshire, 600 feet above the sea, the winter of 1920-21 was severe, keeping the plants of this species asleep until the spring rigours were past, when they put forth their new growth to bear flowers in the following spring. In 1922 they were loaded with bloom, all my plants remaining flowerless. The kind of season, therefore, that an enthusiast in rhododendrons should pray for, is one of prolonged moderate winter cold, followed by a genial spring. Howbeit, even if these conditions are reversed, one may have a splendid display of blossom before evil overtakes the plants in April.

The flowers of different species of rhododendron vary considerably in their resistance to frost-bite. A good opportunity for observing this presented itself a few days before these lines were written. The winter of 1922-3 was the mildest in my recollection: none of our plants had suffered the slightest check, until on 24th March the sun rose upon a very wintry scene. I looked forth in the morning upon a lawn thickly sheeted with hoar-frost under a cloudless sky, and mused gloomily while shaving on the ruin that must have been wrought on the blossom that had made garden and woodland so gay. All this finery I expected to find turned to ashes, nor did I derive any comfort from the bush first inspected—*R. barbatum*—ten feet high, set with hundreds of trusses. Every flower in these trusses was broidered with rime, enhancing their beauty, but only waiting for the sun to

Flowers

strike them to be turned to pulp. But when the sun had topped the hill and lit up the blood-red blossom, the rime disappeared, leaving the flowers untarnished. A thermometer four feet above the grass registered only three degrees of frost, but it must have been much colder on the ground, because on another *R. barbatum* which carried some expanded trusses close to the ground, these were blackened and destroyed, whereas all those higher up escaped injury. On three Chinese species, *R. hæmatochilon, oreodoxa* and *Fargesi*, the blossom remained unsullied; also the Indian *R. campanulatum* and *Thomsoni* (with several of Mr. Gill's hybrids of the latter) had escaped scot free. In fact, the only rhododendrons disfigured by this March frost were *R. arboreum* and *ciliatum*, whereof the flowers are unable to resist even three degrees of cold.

* * * * * * *

Thus far had I written in the early days of April: instead of tearing up the sheets, it may be more profitable to record the chastening experience of the close of that month. We had read about fifteen degrees of frost at Kew, and expressed ourselves arrogantly satisfied with the cloudy clemency of our west coast climate. We ought to have been better prepared for its treachery. During the night of April 25-26, the beneficent cloud-canopy cleared away, allowing the warmth which had accumulated under it to radiate into starlit space, and in the morning the mercury was found to have fallen to 25° Fahr., registering seven degrees of frost. Nearly all blossom on rhododen-

Some Rhododendrons

drons had been destroyed. Well, we had had a feast of that; but the doleful fact had to be faced that much young growth had been blackened and destroyed. Among those that suffered most were *Rhododendron Fortunei, pachytrichum, strigillosum, Davidi, hæmatochilon, Kewense hybrid, lutescens* and the more precocious forms of *decorum*—a species that varies considerably in its period of growth. Species that start their new growth late were none the worse, such as *R. arboreum, Aucklandi, barbatum, sino-grande, calophytum, campanulatum, Fargesi, Soulei, Sutchuenense, Thomsoni, Edgeworthi, Scottianum, crassum, ambiguum*, etc. Also *R. Augustini* and *Schlippenbachii* came through the ordeal with flying colours.

Among other shrubs that happened to be in bloom before this frost it may be noted that *Exochorda grandiflora, Spiræa arguta, Piptanthus Nepalensis, Olearia nitida* and *stellulata* in the open, and *Prostanthera rotundifolia, Clianthus puniceus* and *Calceolaria violacea* on or under walls, never turned a hair. *Cornus Kousa* was badly cut and some of the young growth of *Lilium Sargentiæ* was frosted; but the only herbs that had their flowers destroyed were *Saxifraga Delavayi* and *Beesiana*, the two finest species of the *Megasea* group, except *S. Stracheyi*, which we have had to write off as unable to bear the slightest ground frost.

So endeth the spring lesson of 1923.

I have mentioned shelter from wind as a cardinal requirement of Asiatic rhododendrons; another equally essential, is moisture. Although among their native woods and mountains they can never have a

Flowers

glimpse of the ocean, yet in this country proximity to the sea ensures a humid atmosphere such as they delight in. All the large-leaved species, indeed all except the moorland dwarf rhododendrons, are naturally woodland plants, subject throughout the growing season to frequent mist, drenching rain and an atmosphere constantly and heavily charged with vapour. Open glades in woodland afford the ideal position for them, if with a northerly exposure so much the better, if sheltered from wind. In planting, whether a considerable collection or a few specimens, it is well to act on the advice given me many years ago by the owner of a famous collection in Cornwall—" Place each plant so that you may be able to ride round it in thirty years' time." This is no extravagant counsel. In dealing with the more vigorous kinds it is really the only principle that will enable one to avoid perplexity and disappointment when the plants reach maturity. It may seem absurd to allot a free space of sixty feet in diameter to a shrub only two feet high; but if one grudges it—if one sets another choice species within thirty feet of it—the day will come when one or other must be sacrificed, else they will ruin each other's symmetry. Consider, as a case in point, a noble specimen of the Indian *R. campanulatum* at Leny, in Perthshire. Raised from seed sown in 1823—just one hundred years ago—it is now (or was six years ago) 30 feet high and 150 feet in circumference—a magnificent sight when, in alternate years, it is so loaded with blossom that hardly a green leaf can be seen. Now when this was planted as a small shrub,

PLATE X

Natural size   RHODODENDRON SOULEI   24 May, 1923

Some Rhododendrons

it required no common foresight to estimate and provide for the dimensions it was destined to attain. It may be noted in passing that Leny is situated as near as no matter to the very centre of Scotland, and therefore subject to intense winter cold, yet this and other Indian rhododendrons prosper admirably there.

As an example of the excruciating problems arising from over-crowding choice plants, Lord Stair's collection at Castle Kennedy may be cited. I know of none equal to it in the number and average size of *R. arboreum.* In the middle of the nineteenth century, when these were planted on the advice and with the assistance of Sir Joseph Hooker, nobody foresaw the size to which they would grow. Consequently, they were planted far too close, and now the present lord of that ample demesne, who is enthusiastically devoted to rhododendrons, is confronted with innumerable cases involving the sacrifice of one or more of these huge plants in order to preserve another. It is, indeed, true that rhododendrons, owing to their compact root system, may be successfully transplanted at almost any period of growth; but the labour is very great, and, in these days, beyond the resources of most landowners. I am only expressing the harassing experience of myself and others when I say that the larger and finer Asiatic rhododendrons should never be planted nearer each other than sixty feet, and one hundred feet is not too much. But whereas most of this class of rhododendron are ten, fifteen or twenty years before they make display of blossom, there is no reason why the ground which they are intended ulti-

Flowers

mately to occupy should not be made use of meanwhile. Take the case where a spacious glade has been cut in an old wood, which is far the best kind of provision that can be made for growing the finer species, securing shelter from wind and hot sun, and tending to ward off the dire effects of spring frost. Having set out at ample interval the rhododendrons that are intended ultimately to occupy the glade, there will remain a great deal of bare ground which may be planted with some of the beautiful hybrid rhododendrons which the skill and industry of professional florists and amateurs have provided in such bewildering variety, and which flower freely when of small size. Azaleas also—now classed botanically as rhododendrons—can be had of many charming colours, as well as such shrubs as *Philadelphus, Exochorda, Berberis, Viburnum, Spiræa* and so forth. These will serve to beautify the glade until the principal subjects reach maturity and the flowering stage, when the commoner stuff may be gradually cut out as occasion arises.

The root system of rhododendrons, above referred to, is peculiar. It forms a dense mass immediately under the surface of the ground, never extending beyond the spread of the branches in search of nutriment. In their native forests nutriment is brought to them in abundance from the annual fall of the leaf; but in British woods and shrubberies special means must be taken to protect the roots against evaporation which dries up the soil; against radiation which chills it; against hot sunshine which lowers, and may destroy, root action; and against deficiency of nutri-

Some Rhododendrons

ment. All these four evils may easily be warded off by the annual application of a liberal mulch of withered leaves. It is better to apply the leaves before they dissolve into leaf-mould, because in that condition they permit a free circulation of air, thereby enabling one to lay down a good thick cover, which, gradually decaying, keeps down rank grass and herbage till the bush can screen the ground with its own shade. A layer a foot thick will be found in less than a year to have subsided into wholesome humus an inch deep. To prevent birds scraping away the mulch, sticks, the nearer rotten the better, should be laid over it. Some growers recommend farmyard manure as a mulch for rhododendrons; but I think that material is positively injurious to weak plants, and unnecessary for vigorous ones. All one's concern should be to avoid letting the plants suffer from deficiency of good sound humus; but if an extra stimulus should be resolved on, brewers' draff is both clean and effective.

No idiosyncrasy of rhododendrons is more generally recognised than their inability to thrive, or even to live, in a soil containing more than faint traces of lime. I have in mind an affluent establishment in the south of England whereof the owner was most anxious to cultivate rhododendrons, and resolved not to be baffled by the fact that his demesne was on the chalk. Great beds were excavated at much expense, filled with peaty soil and planted with hybrid rhododendrons. These flourished and flowered well enough until, after two or three years, water charged with lime from the surrounding chalk filtered into the beds;

Flowers

the foliage turned yellow, wilted and ultimately nearly all the plants died.

"Home-staying youths have mostly homely wits"—

and elderly people are not always exempt from a similar disability; wherefore, when Mr. E. H. Wilson first reported having found rhododendrons growing upon limestone in Yunnan, some of us home-stayers were inclined to doubt whether the roots of these plants were actually in contact with calcareous rock, or were merely growing in pockets and fissures filled with humus. But Mr. Wilson is far too well experienced as a botanical collector to make loose statements. His observation in this matter has been amply confirmed by subsequent investigation. Mr. Forrest found many species of rhododendron growing, not only in pockets and crevices of magnesian limestone in North-western Yunnan, but flourishing in the limy talus at the base of cliffs and in soil full of limestone rubble. Thinking it possible that the limestone of this region may differ in some respects from that most commonly found in Europe, or that it may be less soluble, he brought home with him in 1921 samples of the rock and soil in which he found rhododendrons rooted, and we await with interest a report upon them.

Meanwhile, Mr. Forrest's latest researches have thrown suggestive light upon the antipathy to lime shown by every European and American species of rhododendron except the Swiss *R. hirsutum*. It has long been known to botanists that plants of the Heath Order, to which rhododendrons belong, have their

Some Rhododendrons

roots infested by a certain fungus. Now foresters, farmers and gardeners are accustomed to associate fungus in general with mischievous results upon the health of their crops; and in fact many species of fungus do injure and destroy higher forms of vegetation. One has only to mention *Agaricus melleus* that attacks the roots of young Scots pine; the rust *Puccinea* that destroys wheat in the ear, and the deadly *Botrytis cinerea* which causes so much disappointment to the amateur in lilies. But there are beneficent fungi as well as injurious ones, and among them is the fungus which forms the mycorrhiza on the roots of rhododendrons. No doubt it is in some degree parasitic, deriving nutriment from the rhododendron; but it is also symbiotic, amply indemnifying its host by extracting nitrogen from the soil and conveying it to the rhododendron, which is incapable of obtaining that essential food for itself. It seems to follow from this that the rhododendron itself has no direct antipathy to lime in the soil, but that the fungus perishes when brought into contact with lime; and the rhododendron, being unable to draw its own supply of nitrogen, dies of starvation when deprived of the services of its humble messmate.

How, then, can this fungus survive and fulfil its function to the rhododendron in a soil full of lime? Mr. Forrest has found that the under-surface of the leaves of rhododendrons growing on limestone not only harbours the fungus, but that the mycelium penetrates the tissues of the leaves just as it infiltrates and penetrates the roots of rhododendrons growing in

Flowers

soil free from lime. It is true that the fungus has been detected in the felt covering the under side of the leaves of certain Himalayan species of rhododendron which will not suffer contact with lime or chalk; but I am not aware that the mycelium penetrated the leaves of these species. The late Sir Isaac Bayley Balfour satisfied himself that the leaves of rhododendrons growing on limestone in China were so penetrated, and indicated that further research might prove that the fungus, unable to exist in a limy soil, had migrated to the leaves, where it would be able to supply its host with free nitrogen from the atmosphere instead of fixed nitrogen from the soil. This, however, remains at present in the stage of suggestion, awaiting examination and experiment.

It may be noted that *Rhododendron hirsutum* is the only European species which grows naturally on calcareous soil; but its leaves, being destitute of indumentum, are not favourable to harbouring a fungus.

As to the most suitable soil for rhododendrons, it is well known how kindly they take to good, sweet sandy peat, although most of the finest Asiatic species never become acquainted with peat till they are brought to this country. They subsist on forest soil, which is practically leaf-mould, the accumulation of centuries, or in pockets of humus gathered in crevices of the rocks. At Caerhays in Cornwall, where Mr. J. C. Williams has the most extensive collection of rhododendrons in Great Britain—probably in the world— consisting of hundreds of magnificent mature specimens

Some Rhododendrons

and of thousands which will be of fine stature in course of time, there is no peat, and the plants never get so much as a smell of it. Some leaf-mould and sharp sand is put in with them when they are planted out; after which they take kindly to the native loam, liberally mulched with dead leaves, as described above. Nevertheless, peat of the right sort is an admirable ingredient in a compost, both for rhododendrons and many other flowering shrubs; and it possesses this great advantage over leaf-mould that it does not harbour slugs, click beetles and other injurious forms of life which delight to establish their nurseries in leaf-mould. But it must be the right sort of peat, or, rather, peat in proper condition. It must not be used freshly dug out of a moss, for it is then heavily charged with humic acid; but it should be exposed to sun and air for at least six months, well pulverised, and mixed with one-third of its bulk of granite or quartz sand, when it is every bit as wholesome as the thin turves cut off the Bagshot gravel. The sand should be sharp and rather coarse. Sea sand varies much in composition. On some shores it consists largely of triturated shells, containing, therefore, too much lime to make it suitable for rhododendrons; but on other parts of the coast the sand is chiefly composed of particles of quartz and exerts no chemical effect on vegetation. It is easy to test the composition of sand, or any other form of soil, for the presence of lime by dropping upon it a little hydrochloric or sulphuric acid. If this causes effervescence, lime is present; if no effervescence follows, the soil is free of lime.

Flowers

In this country we have not hitherto made enough of the deciduous rhododendrons popularly and conveniently known as azaleas. The group has been recently and thoroughly investigated by Mr. E. H. Wilson and Mr. Alfred Rehder, and the result published under the auspices of Professor Sargent and the Arnold Arboretum in Massachusetts, in the form of a *Monograph of Azaleas*. The price of the volume is five dollars, and is well worth more to anybody interested in that brilliant class of shrub, which not only bears a profusion of blossom in spring but lights up splendidly with autumnal tints on the foliage. In some respects they are more easily cultivated than evergreen rhododendrons, depending less upon a humid atmosphere and being more tolerant of drought.

We owe much of the charm of English park scenery to the skill and judgment with which planting was done by landowners and landscape gardeners in the eighteenth century. Many a piece of flat land, devoid of any natural features to relieve its monotony has been converted into scenes of perennial beauty and interest. Yet those who designed and laid out these grounds had but a small fraction of the material which is now at our disposal. For evergreens they had to rely on Portugal and cherry laurel, holm oak and our native holly, box and yew, with a few conifers. There was no rhododendron until *R. ponticum* and the yellow azalea were introduced in 1768. *R. Caucasicum* followed in 1803, the North American *R. catawbiense* in 1809, and were promptly pounced upon by florists for the manufacture of hybrids; but it was not until

Some Rhododendrons

the advent of *R. arboreum* in 1817 that it became possible to produce the crimson and scarlet varieties that now light up our parks and pleasure-grounds. Those who—like the present writer—temper (some will say dilute) horticulture with botany, may derive more satisfaction from the natural species than from the fairest hybrids; but we cannot remain indifferent to the beauty resulting from many of the crosses that have been effected. It would be sheer pedantry to withhold praise from such magnificent things as *R. Loderi* and *Kewense* with their lily-like blossoms; from the scarlet flames kindled at Tremough by Mr. Gill crossing *R. barbatum* and *Thomsoni*; from the exquisite flowers produced at Caerhays by the marriage of *R. cinnabarinum Roylei* with *R. Maddeni*; or from such old and tried favourites as Lady Eleanor Cathcart, clearest rose so well set off by the deep maroon splash; Broughtoni, with its avalanche of rich carmine trusses; Minnie, with her wanton tawny mottling; Loder's White, which some good judges pronounce to be the finest hybrid ever raised; Lady Clementine Mitford, who loads herself with globes of peach pink; Kate Waterer, lit up by a soft yellow flame in each of her rosy bells; Beauty of Bagshot, with a fine splash of chocolate and red at the back of her white corolla; Ascot Brilliant, whole blood-red; George Hardy, pure white; Sappho, white, with a startling black-purple blotch, and many others, each of which would be prized as a marvel of beauty had we not been so sated with variety. Moreover, the illicit offspring of tender species crossed with

Flowers

a hardier one have often turned out capable of enduring cold which would be fatal to one or other parent, or to both of them. Pink Pearl is a case in point; its pedigree has not been recorded; but if, as seems almost certain, one parent was the Indian *R. Aucklandi*,[1] here you have a splendid shrub that thrives in the coldest districts where *R. Aucklandi* could not be kept alive.

There is only one drawback to planting these hardy hybrids, and it is this. They are nearly always propagated by grafting or in-arching on *R. ponticum* as a stock, which is a most vigorous and persistent grower sure to assert itself unless most sedulously watched. Even the utmost vigilance is often eluded; a long shoot steals up through the heart of a sturdy hybrid unobserved till it betrays its presence by a defiant cluster of reddish-purple flowers. My attention has sometimes been called to this by persons who seemed pleased by what they considered a pleasing freak of nature. The process is going on all over the country; it is quite a common thing to see a plantation of hybrid rhododendrons gradually changing to one of *R. ponticum*, some of the grafts still struggling for survival, others having been entirely suppressed. One firm of nurserymen known to me—there may be others unknown—undertakes to supply hybrid rhododendrons grown from layers. Personally, I would prefer to pay a guinea for a plant on its own roots rather than half-a-crown for a grafted one.

I think I must have said enough to prove that,

[1] The other was probably some highly-coloured hybrid.

Some Rhododendrons

while my heart goes out more readily to the natural species than to hybrids, I am neither insensible of, nor ungrateful for, the skilful florists who have wrought such superb effect upon our gardens and landscapes in general by supplying such abundant decorative material. There is in the natural species a subtle harmony between the hues of foliage and flower which is sometimes missing in highly-coloured hybrids, which are apt to be deficient in the indefinable look of " race " such as distinguishes a thoroughbred horse among a crowd of cocktails. Nevertheless, without hybrid rhododendrons it must be owned that we should be very much more poorly off than we are. Howbeit, it must be also owned that the prospect before us is not a little bewildering. Hundreds of new species of rhododendron have been landed in this country since the beginning of the century; scores of fertile brains and scores of pairs of trained hands are busy with camel's hair brush and muslin bags in the endeavour to go one better than Nature. There seems to be—there probably is—no finality in the process. The enthusiast may ask—why should there be?

" I am not alone in the opinion," says Mr. W. Watson, the late curator of Kew Gardens, " that *Rhododendron ponticum*, when happily situated in a wood or as a large bold mass on the grass in the open, is, when in flower, the most effective of all rhododendrons." [1] Agreed that this species has proved its absolute hardiness and irrepressible vigour beyond all others in this country, and that, summer after summer,

---
[1] *Rhododendrons* (Present Day Gardening Series), p. 3.

Flowers

it varies little in profusion of flower, it must be owned that indiscriminate planting has done a good deal to vulgarise it. Proper care has not been taken to propagate only the better colour varieties, whereof some are admirable, especially the rose-tinted flowers with an orange flame, which stand comparison successfully with many of the Chinese novelties that we treat with so much care. But there is also a strain of *R. ponticum* with flowers of a distinctly disagreeable hue. Within sight of my library window stands a row of large bushes which in every recurring June sets my teeth on edge by flaunting sheets of cold reddish-violet, yet they may not be removed because they protect such precious things as *Eucryphia, Tricuspidaria* and *Rhododendron Soulei* from the cold north-westerly draught. Unlike most of the genus, *R. ponticum* delights in full sunshine and the best effect is never obtained from it in woodland shade, where it shows a tendency to an objectionable " blae " colour in the flowers. But give it a fair chance, and this kindly shrub proves itself well worthy of Mr. Watson's encomium. Such chance has been afforded it at Mochrum Castle in Wigtownshire, the ancient stronghold of the family of Dunbar, now the property of the Marquess of Bute. The castle was a roofless ruin, standing among the Seven Lochs of Mochrum on a vast flat moorland, mostly peat bog. The late marquess took a fancy to the place, restored the ruin and planted a good deal of wood; but for a mile or so on the north and east of the castle there is a stretch of boggy ground unfit for forestry, broken here and there by hog-backed hillocks which were planted

Some Rhododendrons

more than fifty years ago with *R. ponticum*. By good luck a good colour-variety was used; these hillocks are now solid masses of rhododendron, and the result is to show the distant effect which that plant can produce when rightly used. The flush of these bright rose little mountains in June rising out of the flat brown moss is such as I have never seen equalled elsewhere.

Reverting for a moment to Asiatic rhododendrons, it is interesting to note the behaviour of different species in protecting their foliage from injury by frost. *R. sino-grande*, which normally holds its great leaves—fifteen to twenty inches long—like crinkled green morocco leather, does not fold them under the influence of frost, but hangs them vertically. *R. Falconeri*, another large-leaved species, acts in the same way; but *R. Hodgsoni* and *R. barbatum* roll their leaves backwards into tight cylinders like green pencils, giving the bushes a most singular appearance. *R. Thomsoni* depresses its leaves and curls them slightly backwards, but *R. arboreum* advances the outer margins. *R. Smirnovi, Caucasicum* and *pachytrichum* depress their leaves uncurled. Among the smaller-leaved species, *R. cinnabarinum* behaves like *R. barbatum*; but *R. Indicum, ciliatum, glaucum,* and the European *hirsutum* and *ferrugineum* do not alter either the form or position of their leaves even with the mercury standing at 18° Fahr.

Pieris formosa meets frost by depressing its leaves and curling the outer margins forward; while *Myrtus luma* behaves queerly; setting its leaves upright and contracting their margins.

Flowers

In autumn the glistening sticky flower-buds and growth-buds of *Rhododendron barbatum* are thickly covered with corpses of flies of various kinds. I wish that some one skilled in the use of a microscope (which I am not) would hold an inquest upon the dead, in order, if possible, to ascertain whether the slaughter is purposeless or of some advantage to the plant. Some years ago I propounded this question to a good authority on vegetable physiology, and was told that there was no more than a superficial analogy between the capture of flies by this rhododendron and by such plants as *Drosera* and *Pinguicula*; that the strong bristles on the petioles (which certainly hinder an insect's efforts to get free) were not glandular, and therefore incapable of conveying nutrition to the plant. He referred me to the shining glutinous buds of the horse-chestnut as being strictly analogous to those of *R. barbatum*. Against this it may be argued that horse-chestnut buds do not get sticky or shine until the fall of the leaf, when there are but few flies about; whereas those of the bearded rhododendron are both sticky and shiny from their formation in early summer. Moreover, the bristles of this rhododendron are distinctly glandular when first they appear on the new growth, as any one may see for himself by applying an ordinary lens to the young shoot. At what precise period the glands cease to function I have not ascertained; but we shall never arrive at the truth in this matter if we start by assuming that the bristles of *R. barbatum* are not glandular.

I have never seen insects caught on the sticky buds

Some Rhododendrons

of horse-chestnut, though I cannot affirm that this never takes place; but these buds do not become conspicuous till after the fall of the leaf, when there are not many flying insects abroad. When I stood lately before a thirty-year-old bush of *R. barbatum* nine feet in height, and saw every bud thickly plastered with dead flies and here and there a dead spider, which had probably ventured thither to pasture on the flies, the ejaculation of the disciples about the alabaster box of ointment rose to my lips—" To what purpose is this waste ? " Has it no purpose ? Does the plant derive no benefit from such a holocaust ?

Some other species of rhododendron entrap flies, but none that I know of so systematically as *R. barbatum*. The Chinese *R. habrotichum*, i.e. " soft-haired," is of near kin to the Indian *R. barbatum*, and its young growth and leaf-stalks are densely clothed with bristles; but its buds are not shiny nor are they so sticky as those of the other plant, and the bristles are so soft that it only succeeds in catching a few midges. I have seen a number of insects adhering involuntarily to the young shoots of *R. Griffithianum*, which are set with glandular hairs.

Rhododendron enthusiasts enjoy an advantage over those whose passion is for other flowering plants. He whose fancy is for narcissus or tulips may feast to satiety in the early months, but must fast through the rest of the year. The rose-cult commands more disciples than any other, but spring has fled before it responds to their worship; whereas the finer species

Flowers

of rhododendrons exact admiring attention at all seasons. There are but few weeks in the year when one kind or another is not in flower; indeed, *R. auriculatum, discolor* and some forms of *decorum* have already bridged the gap between the July-flowering *R. maximum, pholidotum, micranthum* and *viscosum*, and the earliest flowers of the best form of *R. Nobleanum*, which often open in October and continue through the winter. Some species, notably *R. hæmatodes*, with blood-red flowers, and *R. hippophæoides* with lavender ones, habitually produce a second bloom in autumn. But it is not only by their bloom that rhododendrons fascinate one; the period of growth affords extraordinary variety of beauty and interest. The young leaves of *R. Falconeri* resemble tawny peau-de-Suede; those of *R. sinograndei* are silvery shot with bronze; those of *R. æruginosum* have a bloom like oxydised copper, while the scarlet bracts that accompany the new growth of *R. Thomsoni eximium* and *decorum* are as bright as many flowers.

The genus has proved hitherto remarkably free from disease and exempt from insect attack, at least in the north and west. In very recent years, however, somewhat serious ravage has been wrought in Surrey, Sussex and other southern counties by a hemipterous bug—*Stephanitis rhododendri*—an unwelcome immigrant from the eastern states of North America. " The adult insect is a pretty creature, with a black body and large, yellowish-white, slightly iridescent wings, which are nearly transparent and have clearly-defined

Some Rhododendrons

net-like veins."[1] Eggs are laid on the under side of a leaf, producing grubs of a whitish colour at first, changing later to green with dark brown spines. They remain grouped together, sucking the sap. Mr. Hoare has often found as many as fifty feeding on a single leaf. The leaves become mottled with yellow, afterwards turning brown and the plants cease to grow. In the summer of 1922 I saw some very bad cases among a large collection of rhododendrons in Sussex, but there was no indication that any species or variety was more liable to attack than another. There were instances where, of two hybrid rhododendrons growing quite near each other, one was badly infested, while the other remained quite clean.

The bug is described as being very inactive in all its phases, even the perfect insect, though furnished with large wings, it had never, at the time of my visit, been seen to fly. This may account for the sporadic character of the attack in the garden aforesaid; it was far from general, bushes here and there being disfigured by the parasite, while others of the same species or variety were untouched. In Mr. Hoare's paper, referred to above, an interesting account is given of carefully-conducted experiments with four different washes, and of the result obtained in a Surrey nursery where the stock was badly infested with the bug. The most effective was found to be 1 lb. of soft, high-grade potash soap to ten gallons of water. It would be prudent to make use of this preventive on

---

[1] Mr. A. H. Hoare in the *Journal of the Royal Horticultural Society* for January, 1923.

Flowers

the earliest appearance of the bug. Hitherto it does not seem to have made its way into the north or west, where it is possible a more humid atmosphere may not be to its liking. The small moth—*Tortrix viridana*—whereof the larvæ so sorely defoliate the oaks in the southern English counties, sometimes descends upon rhododendrons and inflicts grievous damage on the young growth, but that species seldom appears in Scotland, and it is only in Sussex and Berks that I have seen rhododendrons attacked.

V

Wild Gardening

> Who that hath reason and his smell
> Would not among roses and jasmin dwell,
> Rather than all his spirits choak
> With exhalations of dirt and smoak,
> And all th' uncleanness which does drown
> In pestilential clouds a populous town?
> <div style="text-align: right;"><i>Cowley</i> (1618-1667).</div>

WILD gardening is an indefinite term denoting the various degrees and methods of beautifying uncultivated ground, or ground not subject to intensive cultivation, with fine flowers and foliage not indigenous to the district. It is difficult to define where this branch of garden craft begins, forasmuch as there are few, if any, country places where some attempt has not been made to enrich the grounds outside the garden proper with flowering shrubs and herbs, and it is impossible to prescribe a limit to the extent to which such enterprise may be carried, because that depends on the amount of labour that can be applied to it. It may be confined to the partial cultivation and adornment of a single woodland glade or open slope, where pretty constant attention can be

Flowers

given to the repression of bramble, bracken and other native growth and weeds. Space so limited may be fenced with wire netting to keep out ground game; but whereas protection of that kind against wild animals is not truly consistent with *wild* gardening, the following notes apply chiefly to a wider and more liberal scheme.

Fifty-three years ago Mr. W. Robinson, to whose example and vigorously-reiterated precepts British gardens owe—more than to any other influence—their present prevailing variety, published a volume on *The Wild Garden*,[1] in which the theme was treated so exhaustively as would render further observations superfluous, but for two factors in the case—first, the immense, almost overwhelming addition that has been made since 1870 to the choice of plants at our disposal by the introduction of hardy exotic species from all quarters of the earth; and second, the fact that Mr. Robinson refrained from any attempt to discriminate between plants that hares and rabbits devour and those that they leave alone. In his lists of herbs for naturalising there are named such desirable things as the common Christmas rose, which would flourish nowhere more freely than on the floor of an open wood, and many of the vast natural order of Crucifers, all of which rabbits seem to regard as having been, if not created, at all events planted, for their special delectation and nourishment. Now, where rabbits abound—and where is there a demesne, other than suburban, that is not more or less infested by these pestilent

[1] London, John Murray, 1807.

Wild Gardening

rodents?—the choice of plants is limited to those species which are either wholly distasteful to them or which regenerate themselves so freely that a fair proportion of the crop may escape their attention. Howbeit, in treating of such plants it behoves me to comply with the ancient motto of the Drummonds, "Gang warily!" because when, many years ago, I acted according to Cicero's doctrine that it is the part of true friendship both to give and take advice [1] and, with sincere intention of being helpful to fellow-amateurs, published a list of such trees, shrubs and herbs as in our local experience had proved immune from attack by rabbits,[2] I succeeded in severely straining the confidence of some old allies in horticulture and incurring indignation of other people personally unknown to me. For years afterward letters arrived —they have not entirely ceased yet—reproaching me for having declared this or that plant distasteful to rabbits, which the writers' rabbits consumed as fast as they were planted.

The truth is that while the voracity of rabbits is insatiable everywhere, the means of satisfying it vary in different districts. Where ivy abounds on the ground I think it may serve to satisfy their appetite in winter, though I cannot affirm this as the result of personal observation. Sheep, at all events, are very fond of ivy; and in mild districts where there is plenty of undergrowth in the woods, rabbits are less likely to

---

[1] Et monere et moneri proprium est veræ amicitiæ.

[2] *Memories of the Months*, First Series, pp. 92-94; Second Series, pp. 48, 49.

attack exotics than in districts where winter is more severe.

Still, one never can tell. I have never known rhododendrons to be touched by rabbits in our woods, except some of the small-leaved kinds like *R. glaucum, præcox* and such-like ; but no longer ago than yesterday, less than four-and-twenty hours before I am writing this note, and at a distance from where I sit of about fourteen miles by the crow flight, I was shown at Logan a singular exception to this rule. Mr. M'Douall has there a splendid collection of rhododendrons, which flourish on an isthmus between two seas where frost seldom comes and cutting winds are shut off by woods and high ground. Here were some bushes of the blue-flowered *R. Augustini,* six and eight feet high, which had been barked completely round and destroyed by rabbits, and this during the winter of 1922-3, the mildest in living memory.

One vice may be regarded as common to the rabbits of every district, namely, inextinguishable curiosity that prompts them to sample, and if palatable to destroy, anything that is newly planted. They never interfere with wild primroses or cowslips ; but when last autumn we planted out a number of strong clumps of coloured primroses, oxlips and polyanthus, they went for them straight and left not a green leaf. The plants were not killed ; they pushed up plenty of fresh verdure in spring, and there is no fear that they will be eaten down in future. The moral of all this is that in managing a wild garden where rabbits abound—nay, wherever a single pair of

Wild Gardening

rabbits exist—it is advisable to protect freshly-planted things during the first six months.

Now as to the plants for the wild garden. Flowering shrubs having been dealt with in another chapter, only some of the heaths and their cousins the Gaultherias and Vacciniums need be mentioned here. The heaths being chiefly a xerophytic or drought-enduring race are the best of all things for covering a dry bank. The tallest of them, *Erica arborea*, which in favoured districts may rise to any height up to twenty feet, cannot be trusted as hardy except in the south and west of Great Britain and in Ireland; but we have lately become acquainted with a variety of this species from the Spanish mountains capable of facing all the cold it is likely to have in this country. It is known and can be had from nurserymen under the name of *E. arborea alpina*. To what height it may attain we do not yet know as it was only introduced to cultivation in 1899; but it has all the other fine qualities of the type which it excels in the vivid green of its foliage. It is an acquisition of the first merit. The period of flowering in both species is prolonged from February to June, and the delicious fragrance of their white blossoms may be enjoyed many yards to leeward of the bushes. *Erica Mediterranea* (so called on *lucus a non lucendo* principle, for it is a native, not of the Mediterranean region, but of Southern France, Spain and, strangely enough, of county Galway) is the hardiest of the tree heaths and bears plenty of pink flowers with chocolate anthers in spring, not so showy as those of *E. australis*, which

Flowers

is not everybody's plant, but flourishes only in the milder districts. The hardy *E. stricta* begins to flower after its vernal congeners have gone to seed, and is intermediate in stature between the tree heaths and *E. vagans* which flowers in late summer and autumn. This brings us down to the lowlier species, such as the delightful *E. carnea*, whereof one should be sure to secure some of the brighter varieties. Those who remember the wild garden which the late Lord Redesdale carved out of a most unpromising landscape at Batsford will not have forgotten the fine use he made of this heath, spread like a carmine carpet over the broad slopes in spring. The late Canon Ellacombe taught me how to treat it as a border edging by clipping it close when the bloom begins to fade, thereby securing a better display in the following March. Much has been written in praise of *E. Darleyensis*, a hybrid originating in the Darley Dale nurseries between *E. carnea* and *Mediterranea*, but in my judgment it is inferior to both, the flowers being of a dingy pink, and their sole merit is that they begin to open in midwinter. Of the many forms of the common heather—*Calluna vulgaris*—it seems to me that the one named *Alporti* is the only variety superior to the wild type. Its flowers are of a rich crimson. *Gaultheria shallon* is a good woodland evergreen, but it must be allowed plenty of room, for it will grow five feet high and spreads into a dense thicket by underground runners. The Chinese *G. Veitchiana* is of lowlier growth, both species being of quiet beauty in flower and fruit. Other Chinese species recently introduced are well

Wild Gardening

spoken of by those who have seen them in maturity, as I have not. *Vaccinium corymbosum* will grow eight feet high and covers itself with pale pink blossom in May, but its chief merit consists in the brilliant scarlet to which its foliage turns in autumn. *Kalmia latifolia*, a most suitable shrub for the wild garden, having been noticed in another chapter, I shall mention only one more of the Heath order, and it shall be the manzanita—*Arctostephalos manzanita*—an evergreen up to eight feet in height. A native of California, doubt might be felt about its hardiness; but the fact that it flourishes in the Edinburgh Botanic Garden (at the foot of a wall, indeed, but not trained to it) shows that it does not flinch from winter cold. This specimen makes a striking display of chalk-white blossom from January till April. Hitherto this species has been very rare in British gardens, probably owing to difficulty in propagating it from cuttings, and I have searched trade catalogues for it in vain.[1]

All leguminous plants are peculiarly attractive to browsing and gnawing animals and require protection from them when young, but the taller brooms soon acquire woody stems with such a tough bark as enables them to defy attack. The white Spanish broom (*Cytisus albus*) and its hybrid the Warminster broom (*C. præcox*) are objects of such surpassing beauty when they deck their slender sprays in May and June, the

[1] Mr. Stewart, of the Edinburgh Botanic Gardens, has overcome the difficulty that has baffled the attempts of so many persons to cause cuttings of this species to strike. His success in propagating many other plants is very remarkable. He has even raised the common beech from cuttings, which I believe has never been done before.

Flowers

former with milk-white, the latter with sulphur-coloured flowers in lavish abundance, that it is a pity they are not more frequently planted. Both are admirable ornaments of the wild garden.

Our eighteenth century forbears, not so amply provided with choice of evergreens as we are, seem to have planted butcher's broom (*Ruscus aculeatus*) very freely; at all events it is very often to be seen in the grounds about mansions dating from that century. Unluckily they were not careful to plant both male and female of this queerest member of the Lily order, wherefore it is seldom that one sees its sombre foliage lit up with scarlet berries as big as wild cherries, with which it brightens the hedgerows in parts of Hampshire. Foliage, quoth I? that is a misnomer when applied to a plant that is practically leafless. What may easily be mistaken for leaves are really needle-pointed " cladodes " or woody branchlets flattened out to look like leaves, the true leaves appearing aborted as minute dry scales at the base of each cladode. The flowers are very small, of a dingy white and appear on what is really the upper surface of the cladode; but the stem of the cladode is twisted so that the flower, already so inconspicuous as to escape all but careful examination, is rendered still more so by being set on what has become the under side of the cladode. It is by a strange freak of nature that a plant claiming close kinship with such gorgeous things as lilies, tulips and hyacinths should take such elaborate pains to conceal its insignificant blossoms, and that these should be followed by large, brightly-

Wild Gardening

coloured and conspicuous fruit. All the shrubs mentioned above, not excepting the last, require protection against rabbits until they are well established; after which they may be trusted to hold their own. And for the protection of those that are planted out quite small—say six inches to a foot high—let me make grateful advertisement of a device supplied under the name of wire-netting seed-guard, by Messrs. Smith, Fletcher and Co., 172 High Street, Edinburgh. It is a circular cage of wire-netting, supported on a frame of stout wire, the side supports being prolonged into spikes for fixing in the ground. I have found these cages simply invaluable; they can be used for the protection of a long succession of young things, for they are of enduring material.

The surest way to get hardy herbs other than bulbous plants naturalised in what is intended to become a wild garden is to scatter seeds; but to do so indiscriminately is a mere waste of good material. It is the height of futility to sow it upon close turf or on ground covered with other herbage. The surface must be severely scarified to ensure the establishment of a fresh colony, and advantage may be taken of any ground laid bare by the removal of shrubs or undergrowth. Most of those who have to deal with any soil except chalk or limestone have had occasion to clear away thickets of the common *Rhododendron ponticum*, especially where it has been planted too near carriage-drives and woodland walks. On cretaceous soil there is often a dreary congestion of Portugal and cherry laurels, privet, etc. Now, when such things are rooted

Flowers

out, the opportunity for seed-sowing should not be neglected, because the dense growth of rhododendrons and laurels will have destroyed all vegetation beneath them, leaving an admirable seed bed. Charming effects may be secured by scattering in such places seeds of white foxglove, *Campanula lactiflora* and *latifolia,* the willow gentian (*G. asclepiadea*) and several species of mullein (*Verbascum*). All these may be relied on to keep a hold once the chance is given them. The foxglove, of course, is only biennial, but seeds itself in profusion. The willow gentian is naturally a woodland plant, and its rich blue wands are even more attractive in such a situation than in the garden proper. The same may be said of the two great bell-flowers afore-named, which become a positive nuisance in a border, well-nigh impossible to eradicate, but greatly enrich a woodland. Of the two species, our indigenous *C. latifolia* endures deeper shade than the other. The wood forget-me-not, *Myosotis sylvatica,* is apt to take too complete possession of herbaceous borders; but it is a shallow-rooting plant, easily ousted from places where it is not wanted. It is a simple matter to establish it with excellent effect in the wild garden by throwing down the old plants when the seed is ripe on any piece of bare ground, thereby ensuring a mist of sky-blue in the following spring.

The Primula family, including our native primrose and its varieties of many colours, contains many fit subjects for naturalising. It amuses me to remember that in the early 'seventies of last century I paid thirty shillings for a single plant of the newly-introduced

Wild Gardening

P. Japonica—at that time an exciting novelty, for it was the forerunner of a class carrying several tiers of flowers—and to contrast that experience with the impression received (in this very month of June that I am writing) from a visit to the wild garden planned and planted by my friend and neighbour, Mr. Kenneth M'Douall of Logan, through the ample grounds of that demesne. A broad and shady carriage-drive winds through a wood from the flower-garden proper to the sea-shore, a distance of fully a quarter of a mile. Giant rhododendrons—the crimson hybrid which dates back to the introduction of the blood-red *R. arboreum* in 1817—rise to a height of over thirty feet on each side of the way, and were in gorgeous bloom at the time of my visit; great fuchsia bushes were there also, preparing to take up the display when the rhododendrons should be past; but more remarkable than either were the primulas, chiefly *P. Japonica* and *pulverulenta*, which illumined the sides of the whole length of the roadway with all shades of crimson and pink and pure white. The effect of evening sunlight shining through this mass of colour is such as I can never forget. The plants have taken as complete possession of the ground as blue hyacinths have done further within the wood—they sow themselves literally in millions, many of the seedlings being natural hybrids between the two species, and a new note of colour will result when the orange *P. Bulleyana*, the canary-yellow *P. helodoxa* and the bright-eyed *P. Burmannica* have had time to establish themselves. To ensure such a display as this, moist soil and partial shade are

Flowers

essential, conditions which all these robust primulas demand for their just development. *P. pulverulenta* and *helodoxa* rise to a height of between three and four feet when they are suitably placed.

The American wood-lily—*Trillium grandiflorum*—is another lover of semi-shade and moisture, and never shows to such advantage as in the woods. If peat can be found or furnished for it, so much the better; but well-rotted leaf-mould with an admixture of coarse sand answers equally well. Rabbits leave this beautiful plant alone after it has become established, and it may be trusted to produce plenty of seed, but whereas the seed is heavy and lies where it falls among the old plants, it is well to collect it when ripe and scatter it upon any bare, moist space.

Funkia Sieboldi is a good subject for naturalising, for although its flowers come just short of being handsome, the foliage is fine, and is possessed of some quality that renders it absolutely distasteful to ground game. All species of wolf's-bane, monkshood or aconite, as we indifferently name the genus *Aconitum* in English, are equally immune from attack by rabbits, are equally happy in sunshine or shade, and provide among them a long succession of flower beginning with the common *A. napellus*, which hoists its blue ensign towards the end of May. Not a very clear blue, indeed; but there follows a long succession of other species showing brighter colours—*A. Fischeri, Wilsoni, Chinense, Japonicum*, etc. I value the common monkshood for the early assurance it gives of the approach of spring by spreading a vivid green carpet over the

Wild Gardening

dead leaves of bygone summers. *Astilbe Davidi* is an obnoxious spreader in gárden borders; but its tall flower spikes, crimson with a blue sheen, show to admirable advantage in a woodland glade, where it is easily naturalised if the soil is on the moist side.

Our list is already a long one; but before having done with herbaceous things room must be found for mention of *Geranium Ibericum*, royal blue shot with blood; *G. Armenum*, strident magenta tempered by distance to good crimson; *Centaurea macrocephalum* carrying golden globes aloft and *C. Rhaponticum* with rosy ones; *Libertia grandflora* or *formosa* (I really don't know which is the species that sows itself so freely here) with long sprays of pure white; *Spiræa aruncus*, with cream-coloured plumes, *S. palmata* with crimson and *S. lobata* with rose-coloured ones; *Buphthalmum speciosum* and *Inula grandiflora*, or still better *Inula Roylei*, with bold orange discs; lastly, *Lavatera Olbia* which, for three months on end, ceases not to string its graceful wands with great blossoms of clear carmine.

The family of *Anemone* contains many species most suitable for naturalising, and the blue-flowered species *A. Appenina* and *blanda* are often made good use of in that way; but I have never seen advantage taken of the fine *A. alpina* and *sulphurea* for the purpose to which they seem so admirably adapted, namely, filling a sunny glade in the wild garden. They are not shade-lovers; both of them carry their charming flowers to a height of two feet; while *A. alpina* rejoices in a limey soil, *A. sulphurea* refuses it, demanding peat and loam.

Flowers

No suggestion for the furnishing of a wild garden would be worth the paper it is written on without a mention of bulbs, for that is a class of plant which resents more than most others being fidgeted by a fork or harassed by a hoe—implements that have to be plied in any well-regulated herbaceous border. Narcissus, of course, in its numerous species and innumerable varieties, is indispensable, and may be taken for granted without further notice except that all the species and most of the florists' varieties thrive in almost all, if not all, localities, as well in grass as in garden borders. A judicious selection will ensure a display for fully four months. In the present year, 1923, *N. cyclamineus* began to flower on 8th February, and on this day, the 8th of June, there are still sheets of a late-flowering variety of the pheasant-eye—*N. poeticus*—although these have received some disfigurement by cattle, not browsing on them, but lying upon them. Snowdrops, snowflakes and winter aconite may be planted *ad lib.*, and the quiet grace of *Ornithogalum nutans* should secure favour. The common fritillary—*F. meleagris*—both coloured and white—soon form strong colonies, self-sown in places where rabbits come not and where the soil is not too dry. The Pyrenæan species—*F. Pyrenaica*—with flowers of perfect form, sub-fusc with gleams of gilding on the lip, requires a little more coaxing, but repays it manifold. The squills lend themselves generously to the adornment of the wild garden, one species—*Scilla nutans*, which English people vex their Scottish compatriots by calling blue-bells—being very likely to

Wild Gardening

invade it without waiting for an invitation. But whereas one should aim at giving the wild garden some character to distinguish it from the ordinary run of woodland, delightful as that undoubtedly is in spring, it is well to anticipate the coming of the blue hyacinth by something less familiar. To this end there is no bulb that lends itself more readily than its Spanish counterpart, *S. Hispanica* (*campanulata*) whereof there are several varieties—pure white, blue in various shades, and flesh colour—all equally hardy and easy to establish. The blossoms of this species are more open — more truly bell-shaped — than those of the British wilding. The flowerscape is more substantial and the flowering season somewhat later.

Among spring-flowering bulbs of lowlier stature the common dog-tooth violet—*Erythronium dens-canis*—makes itself as comfortable on the margin of an English wood or shrubbery as I have seen it doing on the bleak limestone uplands of Montenegro, where the goats devour every green thing indiscriminately that is not on ledges inaccessible even to them. In this country, however, rabbits leave the dog-tooth's beautifully mottled leaves severely alone, but pheasants exasperate one by nipping off the rosy blooms. Some of the American species of *Erythronium* are as easy to grow as the common dog-tooth and make much more show. Mr. Carl Purdy contributed a useful key to the western species (*Flora and Sylva*, vol. ii. pp. 250-256), and the liking of this genus and many other genera of garden plants

Flowers

for wood ashes is well noted in the following observation by him :

"During our dry summers brush or forest fires are common. Before a fire I have often seen *Erythronium californicum* growing in brush lands to a height of six or eight inches with but a single flower; after a fire it may be as much as sixteen inches or two feet high, with from four to sixteen flowers."

The loveliest species of this interesting genus is *E. revolutum*, especially the variety *Johnsoni*, with bright rose flowers and the flesh-coloured one called "Pink Beauty." These, as well as most of the other species, are regularly offered for sale by the leading nurserymen in this country, but few amateurs seem to take advantage of the offer.

Among tulips the yellow-flowered, fragrant *Tulipa sylvestris* is recognised by Bentham and Hooker as indigenous to some of the eastern and southern counties of England and naturalised in others. We have not tried it yet outside the garden, but the freedom with which it has spread through a bed of roses from an original planting of half-a-dozen bulbs seems a fair indication of its probable behaviour in the wild garden. Having seen many hundreds of tulip bulbs of garden varieties planted in orchards and other grassy places without permanent success, I had formed the opinion that when so treated they seldom endure more than a season or two ; but a few weeks before writing these notes, in perambulating the grounds of Mr. J. C. Williams, at Caerhays Castle in Cornwall, I came upon a grassy glade peopled with scores of tulips, apparently of the cottage or Darwin variety. They

Wild Gardening

were just about to flower, and a week or ten days later must have made a striking display. I was told that these bulbs had been planted some forty years ago, so the experiment may be worth trying elsewhere; but I must own that although I have seen it tried in many places, the result has not been encouraging.

There are many species of iris that may be planted in the wild. The bearded and other rhizomatous species mostly thrive to perfection where they can be well roasted in sunshine; but the bulbous section are more accommodating, and the so-called English iris—*I. Xiphium*—which is not English at all, but Pyrenæan—in all its splendid range of colour may be used with excellent effect beside woodland paths. Unluckily, rabbits esteem it a special delicacy.

Of the true lilies there are four species which our experience here enables me to warrant with confidence as good subjects for the woodland or meadow ground. There may be more, but I have not made experiment with them. *Lilium giganteum* never shows so well as among natural surroundings. Plant each bulb singly in deep soil as rich as may be and in a position well sheltered from the wind. You may have to wait two or three years before the great flowering stem ascends, but after that the single bulb will have surrounded itself with a number of others, from among which you may expect an annual display. The panther lily—*L. pardalinum*—revels in a cool, grassy glade and so does *L. martagon*—the beautiful white variety as well as the coloured type. As for the common Turk's-cap—*L. Pyrenaicum*—one can hardly

Flowers

put it wrong, so easily does it adapt itself to almost any kind of soil and degree of exposure. I have seen it in the deep shade of a thick wood, on a railway embankment over a windswept moorland, in a sun-baked border, and in the snug recess of a cottage wayside garden, in all these situations a picture of health and content. Fastidious persons complain that the scent of this kindly lily is disagreeable. They can avoid that by keeping to windward of it; to me it is agreeably reminiscent of the days that are no more. I might add a fifth species to the list, for although we have not planted *L. Davuricum* var. *incomparabile* in the wild, I have in mind a splendid conflagration of this species upon a little island in the Itchen, not far below Winchester.

It is time this rambling chapter were brought to a close, which may most profitably be accomplished with a warning. During the last thirty or forty years, more new species of trees, shrubs and herbs have been introduced to this country than in any preceding similar space of time. We amateurs have been apt to lose our heads in the excitement of testing novelties, and want of discrimination has landed some of us in trouble. The genus *Polygonum*—the knotweeds—contains a great number of beautiful flowering plants, ranging in height from the thirty-foot climber *P. Baldschuanicum* to the creeping *P. vaccinifolium*, both of them herbs of much merit. Accordingly, when the Asiatic *P. Sachalinense* and *polystachyum* were placed at our disposal by the enterprise of collectors in Asia, we chose the most favoured spots for them and planted

Wild Gardening

them in the choicest company. How eagerly did we watch the shoots that appeared in the following spring, and when September came round how proudly we invited our friends to admire the great flapping leaves of *P. Sachalinense* on its eight feet stems, and the fragrant sprays with which, like delicate lace-work, *P. polystachyum* veils its arching wands. Little did we suspect what was going on underground—long, tough, rope-like roots boring their way in all directions, only betrayed their meandering in the following year, when they sent up suckers, rivals in vigour of the parent plants, many yards away from their source. It was an ill chance that caused us to plant the terrible *P. polystachyum* just outside a wire fence protecting a flower border. Promptly it burrowed under it, got into the border, whence in the course of twelve years we have failed to get rid of it. It is a charming thing in a woodland glade, far from any cultivated ground, for it puts forth its lovely flowers in September and fills the air with delicious scent till the first frosty morning puts an end to it. *Polygonum alpinum* in June and *P. campanulatum* in August, are both pretty things and inoffensive in the matter of rambling, for they do not send out runners underground. All these knotweeds have a succulent appearance, but they possess some principle that renders them utterly distasteful to rabbits. *Polygonum* is a most Protean genus, the extravagant variety of forms which it assumes is infinite. I sometimes puzzle visitors by taking them to a plant of *P. equisetiforme* and asking them to determine its affinities. It came here from

Flowers

Kew some years ago, but I do not know its country of origin. It is quite unlike any other knotweed, having two-foot rushlike stems of dark green, with a few minute clasping leaves near the base of them, and in September these stems become wreathed with small white, star-like, sessile flowers—quite a pretty plant.

VI

The Choice of Plants

> For lo ! the winter is past,
> The rain is over and gone ;
> The flowers appear on the earth ;
> The voice of the turtle is heard in our land.
> *The Song of Songs*, ii. 11, 12.

IT was all very well for Martin Tupper to describe memory as "the storehouse of the mind, garner of facts and fancies"; but a storehouse is of little use unless its contents are so arranged that one can lay a hand upon any article whenever occasion arises for using it. Nobody can have given much attention to garden craft without finding it indispensable to keep some kind of record of the character and behaviour of the plants that have pleased or disappointed him; and whereas no private collection can contain more than a small fraction of the plants that may be grown in the open air of this country, it is chiefly upon notes made in his own garden or the gardens of others that one must rely for guidance in the choice of species. A note-book and pencil, therefore, should always be in one's pocket in visiting

Flowers

botanical gardens, private gardens or nursery grounds. It is to such note-books that I turn before venturing to offer assistance in discriminating between more and less desirable flowering plants. The better to explain my purpose, a leaf from one of these note-books may be printed here as a sample showing the relative merit of different species of the same genus, in the writer's opinion.

SHRUBS.

Superior.	Inferior.
Senecio Greyi	Senecio compactus
Tricuspidaria lanceolata	Tricuspidaria dependens
Berberis pinnata	Berberis aquifolia
Escallonia × Langleyensis	Escallonia rubra
Osmanthus Delavayi	Osmanthus ilicifolius
Hamamelis mollis	Hamamelis Japonica (arborea)
Viburnum Carlesi	Viburnum crassifolium
Viburnum tomentosum Mariesii	Viburnum utile
Viburnum fragrans	Viburnum rhytidophyllum
Pieris Japonica	Pieres floribunda
Spiræa bracteata	Spiræa × Van Houttei
Veronica cupressoides	Veronica salicornis
Corokia virgata	Corokia cotoneaster
Senecio Hunti	Senecio Munroi
Rhododendron × præcox	Rhododendron mucronulatum
Rhododendron Moupinense	Rhododendron parvifolium
Olearia macrodonta	Olearia Traversi

HERBS.

Superior.	Inferior.
Camassia Cusicki	Camassia Leichtlini
Podophyllum Emodi	Podophyllum peltatum
Cynoglossum nervosum	Lindelofia spectabilis
Veronica longifolia sub-sessilis	Veronica spicata
Veronica Virginica	Veronica ambigua

The Choice of Plants

HERBS—(*continued*).

Superior.	Inferior.
Aster Fremonti	Aster (Erigeron) salsuginosus
Gentiana septemfida	Gentiana cruciata
Gentiana Farreri	Gentiana phlogifolia
Gentiana Sino-ornata	Gentiana ornata
Gentiana Lagodeshiana	Gentiana macrophylla
Aster sub-cœruleus	Aster alpinus
Saxifraga × apiculata	Saxifraga sancta
Saxifraga Burseriana	Saxifraga juniperina
Saxifraga × Red Admiral	Saxifraga " Guildford seedling "
Saxifraga × Bathoniensis	Saxifraga muscoides
Saxifraga (Megasea) Beesiana } Saxifraga (Megasia) Delavayi	Saxifraga (Megasea) crassifolia varieties
Hypericum reptans	Hypericum repens
Adenophora megalantha	Adenophora Lamarckii
Pulmonaria angustifolia	Pulmonaria Arvernensis
Adonis vernalis	Adonis Amurensis
Cynoglossum Appeninum	Anchusa Italica
Pæonia Mlokosewitchii	Pæonia lutea
Papaver umbrosum	Papaver aculeatum
Mimulus × Bartonianus	Mimulus Lewisi

Lists such as these may be extended indefinitely, ever lengthening with the passing seasons. They may be taken as indicating under the head of " Inferior " plants that should not be admitted to select company. Let me now run through my note-books and jot down the names of those herbs which have distinguished themselves as desirable in a flower-garden of moderate dimensions.

In the early months one relies chiefly on flowering shrubs and bulbous plants for display, and these have been discussed in other chapters. Among spring-flowering herbs, those species of the *Megasea* section of

Flowers

saxifrage hitherto commonly grown are lacking in grace, the leaves being too many and gross for the somewhat stodgy panicles of flower, and the more refined *S. Stracheyi* is too tender for outdoor work in all but the very mildest districts. But in the Chinese *S. Delavayi* and *S. Beesiana* we have two very showy plants, holding their rich crimson flower-stems well aloft. The leaves, also, are deeply stained with maroon and red.

Pulmonaria saccharata is, on the whole, the best of the lungworts, for it begins to open its flowers very early in winter (in the present year of grace, 1922, it was out on Christmas Eve) and continues till Easter, after which come the fresh dappled leaves, well-nigh as ornamental as the blossom. Next best is the deep blue form of *P. angustifolia*, which may be given a fitting companion in the variety named *rosea*, with flowers of *rose du Barry*. Of the same natural order is *Anchusa myositidiflora*, which, despite the burden of these polysyllables, throws round itself a shower of bright blue flowers, beginning in February. Unlike most of the family it likes moisture, if combined with good drainage. *Omphalodes Cappadocica* is another borage-wort of recent introduction, said to be a really good thing, but it has not been long enough here to show its quality.

Cardamine trifolia, a European relative of our common Lady's-smock, is a neat and safe carpeting plant, with plenty of cheerful heads of white flower on six-inch stems in March. In the same order of Crucifers may be noted the spiny madwort—*Alyssum spinosum*

The Choice of Plants

—often grown as a rock plant, but very ornamental in the open border, where it has made here immense cushions six feet in diameter, which are very satisfying when densely covered with tiny white flowers in May and June. I believe there is a pink variety, also, but not having met with it, I know not whether it is a good colour, or the faint, uncertain flush that mars the purity of lily-of-the-valley in the variety sold as *rosea*. This *Alyssum* in the course of years is apt to get bare in the centre of the cushion, the old stems dying out. A nice effect may be obtained by dumping a barrow-load of loam and leaf-mould upon it and sewing a few seeds of Shirley poppy or other choice annual; the original plant will then show as a broad band, white when in flower, silvery grey at other times, encircling the gay poppies. As a plant for a retaining wall *A. spinosum* has a fine effect.

Having referred above to lily-of-the-valley, opportunity may be taken to note that the more desirable variety of that most popular plant is that whereof the flowers are followed by scarlet berries. In our woods at Monreith there are broad spreads of lily-of-the-valley, but they are of the barren kind. Not a single berry do these plants ever produce; though my envy has been stirred at seeing them in the gardens of other people.

Under the name of *Campanula amabilis* we have a single plant of a very nice bellflower, very distinct from any other of that genus known to me. I know not its origin, nor can I find notice of it in any trade list or work on gardening, except the late Mr. Farrer's

Flowers

Alpines and Bog Plants, page 140. He describes it as being " pleasant as its name ; a stout rosette, with three foot spires, loose and graceful, of big shallow cups, soft blue with a dark purple eye." In our plant the shallow cups are deep purplish-blue with a lustrous sheen on the petals, and the stems are only fifteen to eighteen inches high and very brittle. The radical leaves are fleshy, with purplish undulate margins. Altogether a plant of distinction, but somewhat chary of flower.

Anemone rivularis, rupicola and *narcissiflora* are summer flowers of merit and of easiest culture, yet you may go through a hundred gardens without meeting one of them. I have spent years in trying to coax the Japanese *Anemonopsis macrophylla*[1] to behave here as it used to do in Mr. Grove's garden at Kentons, flinging its white-fronted, blue-backed flowers freely above the deeply cut leaves. All to no purpose : every spring there is a fine flush of foliage ; then the flower stems push up, and " Ah," thinks I to myself, " see what perseverance combined with technical knowledge will accomplish ! We really have solved the secret of this fickle fair one at last." Next time I pass that way, the foliage is turning yellow, the flower-stems have begun to wilt, and I ejaculate something which will not bear repetition on this chaste page.

The genus *Primula* enriches spring and early summer with a long succession of truly beautiful

---

[1] Not to be confused with *Anemopsis*, a synonym for the California *Houttuynia*, one of the Piperaceæ.

The Choice of Plants

flowers; but whereas a considerable proportion of the very numerous species introduced of late years from the Far East are of biennial duration only, or at least most apt to behave as if they were constitutionally monocarpic, it may be useful to indicate those which are vigorously perennial, and may be trusted to have come to stay when planted in moist soil not too freely exposed to sunshine. *P. denticulata* and the more vigorous *P. Cashmeriana* bear purple or lavender flowers in spherical heads. The latter is said to be the finer plant; we have not tried it here yet; it must be fine indeed to be better than *P. denticulata*, which sometimes sends up its stout flower stems in January, but may be expected at its best—and very good is its best—in March. Then follows a long succession of the kinds that flower in tiers throughout May, beginning with *P. Japonica* in all tints between dark crimson and pure white, followed by *P. pulverulenta*, with a shade of brown in its red corolla and culminating with *P. helodoxa* rearing its canary-yellow whorls sometimes to the height of four feet.[1] About the middle of the month *P. Bulleyana* opens the lowest tier of its red-gold pagoda, after which *P. Burmannica* and *Beesiana* strike in with a note that it would be uncivil to such gladsome flowers to describe as magenta. *P. Poissoni* delays till mid-June or later, flying the same strong colour. Most vivid of all, but much less robust and uncertainly perennial is the fiery orange-

---

[1] The flowering dates are only approximate, taken from a record kept at Monreith for the last twelve years. In some districts the flowering period will be earlier, in others later.

Flowers

scarlet *P. Cockburniana*. If a moist border facing north or west can be given up to these plants, they will flirt assiduously with each other, resulting in hybrid offspring of many degrees of beauty and shades of colour. Here I broke off to walk across the lawn to measure the height of a come-by-chance hybrid between, probably, *P. pulverulenta*—perhaps *P. Japonica*—and *Bulleyana* or *Cockburniana* (the parentage of these natural crosses is very uncertain). The flowers are a charming blend of carmine and apricot, and stand forty-seven inches high in their stocking-soles, so to speak.

Mr. E. A. Bowles's garden books are such a refreshing source of information and provide such a comprehensive calendar of blossom that it is surprising to find in *My Garden in Autumn* no mention of the beautiful family unbeautifully named bugbanes—*Cimicifuga*. Perhaps his borders at Waltham Cross are too hot for them, despite their proximity to the New River, for these plants like a cool soil and plenty of moisture. Stranger still, Reginald Farrer, whose feeling for flowers was pretty catholic, does something more than damn them with faint praise, pronouncing them

"too unmistakably ominous in appearance to have the full attractions of their beauty... It is fortunate that poison-plants seldom if ever fail to offer some indication of their nature to the sensitive observer. Who would take *Atropa*, *Hyoscyamus*, or fatal hellebore for ordinary benign, innocent creatures; or fail to feel an intuition of evil about monkshood, foxglove and Daphne mezereon?"[1]

---

[1] *Alpines and Bog-plants*, page 188.

The Choice of Plants

It was seldom that this gifted writer allowed his pen to ramble into such nonsense. What sinister suggestion is there in flower or leaf of the Christmas rose (*Helleborus*) ? and who so sensitive as to shrink from inhaling the early fragrance of mezereon? Why, some of the many species of rhododendron which Farrer described in such glowing terms and in pursuit of which he laid down his life, are deadly poisonous. The younger Hooker lost several of his mules that perished after browsing on *Rhododendron cinnabarinum*. No; bugbanes did not appeal to Farrer as they do to me, who would sorely miss them if they were not present to gladden the autumnal borders. *Cimicifuga simplex* latest and, after the American black snake-root—*C. racemosa*, perhaps best of the family, carries its snowy plumes to a height of four feet, the petals and protruding anthers being pure white and the sepals daintily touched with carmine. Other species are *C. Japonica, racemosa, fœtida, cordifolia, Dahurica* and *Americana*, all beautiful flowers—it is hard to say which is the fairest, and all distinguished by their evil stink, which has earned for them the name of bugbane on the homœopathic principle of "like cures like," the flowers being dried by the natives of Siberia and used to drive away bugs. They are all woodland dwellers, impatient of drought or burning sunshine.

Mimulus Bartonianus is a novelty of much merit, a hybrid between *M. cardinalis* and *Lewisi* raised by Major Barton in his Irish garden. It is of shapely, erect habit, carrying a long succession of crimson

Flowers

flowers to a height of eighteen inches or two feet. As it forms no seed, it must be propagated by cuttings, which strike readily. Good scarlet forms of one of its parents, *M. cardinalis*, are excellent, but there are also some of impure colour which should be suppressed. The other parent, *M. Lewisi* we have discarded as not worth growing. *M. glutinosus*, better known by the name of *Diplacus*, is usually treated as a greenhouse plant, but can be trusted in the open, under a wall for choice, in mild districts. Its flowers are of a soft apricot hue and beautifully formed.

Mimulus radicans only requires to be given a chance to form an ideal carpet in the rock garden.[1] It is a native of New Zealand, and is neither aggressive like *Erigeron mucronatus* nor asphyxiating like *Arenaria Balearica*. Creeping low and close, it covers the ground with spoon-shaped inch-long leaves, yellowish olive mottled with bronze. Towards the end of May, the carpet becomes spangled with gay flowers, immense as compared with so lowly a plant, the lower lip three-lobed, shell-white with a bright yellow patch; the upper lip two-lobed, rich violet and erect. Both leaves and flowers are well worth examining under a lens, and the little plant makes itself so well at home as to outrun the border and mingle with turf on the lawn.

[1] I have retained the name given to this plant by Sir J. Hooker, which also is given in the Kew *Hand List*; but Mr. Cheeseman has removed it to the genus *Mazus*, because Hooker described it from imperfect specimens, and it has "the habit, inflorescence and calyx of *Mazus*." Cheeseman describes the flower as white, with a yellow centre, but in the plant we have here, the beauty of the flowers is much enhanced by the rich violet of the upper lip and throat.

The Choice of Plants

Another pretty carpeting plant, equally innocuous to its neighbours, is *Veronica filiformis,* not to be confused with *V. filifolia,* which is many times taller than this gentle creeper, which rears its pretty blue and white flowers not more than an inch or two from the soil. It came as a gift from my good friend Mr. E. A. Bowles, and it flourishes quite as happily in our humid west as it does in his sun-baked garden, flowering throughout the summer.

The genus *Codonopsis,* closely akin to *Campanula,* contains some very interesting species, the best that we have here being *C. ovata* or *sylvestris* (I am uncertain which, or whether these are synonyms), with palest grey-blue bells, painted inside with purple, yellow and black, and the far showier *C. meleagris.* They only require to be planted in well-drained soil, in sunshine or partial shade, to flower freely in June and July. *C. viridiflora* lays no claim to beauty; it is a climbing plant with green, evil-smelling flowers. Like all the bell-flowers and their kin, *Codonopsis* appeals to both rabbits and slugs as irresistibly as chocolate does to schoolgirls.

Having mentioned slugs, I may make it an opportunity for communicating a recipe which a neighbour has lately applied to the destruction of these, perhaps the most destructive vermin that infest gardens. Make up a paste of flour and stale beer (why stale?), spread it on a board and lay the board over-night with the paste downwards in the border. My neighbour tells me that on the first morning he found over one hundred slugs collected under

Flowers

the board. Rest assured that he did not set them free!

Campanula Sarmatica, from the Caucasus, is a species not frequently seen in gardens. It serves us instead of the more refined *C. barbata*, which resents our wet winters by persistently behaving as a biennial, refusing to face a second summer. The flowers of the two species are similar in colour and shape, but *C. Sarmatica* bears them in many one-sided racemes. I brought *C. rhomboidalis* long ago from a garden in Sutherland, where it excited my cupidity by far-seen sheaves of deep-blue bells; but it has never behaved so generously in our lowland latitude.

There is no handsomer species than our native *C. latifolia*, for which, strange to say, Dr. Prior has not recorded any popular name. It grows four feet high, forming a long spike of large flowers in the axils of the leaves, varying in colour from purplish blue to pure white. It sows itself somewhat too readily in borders that are to its liking—moist and rich—but it is one of their chief ornaments in July. It may be easily naturalised in woods, coming on as the foxgloves pass out of flower, and is one of the few showy herbs that flourish in deep shade. I fancy, but am not sure, that, like the foxglove, it does not relish lime in the soil.

The Siberian globe flower—*Trollius patulus*—deserves to be better known and more frequently grown than it seems to be, for the sake of the clear golden discs which it rears on stiff stems nine inches or a foot above nicely-marbled leaves. Unlike most of the genus, its flowers are not globular in shape, but flatter

The Choice of Plants

than those of a king-cup, *Caltha palustris*. Larger in all its parts, but very similar in other respects, was a globe flower that came here under the name of *T. Yunnanensis*, a really splendid thing which, unluckily, has disappeared owing to the place assigned to it having been too dry. Passing an hour of a June day in the ancient, but up-to-date, nursery of Messrs. Cunninghame and Fraser in Edinburgh, I noticed, and promptly secured, two varieties of *Trollius* named " Orange Crest " and " Y. Smith," both most desirable plants. Our native *T. Europœus* deserves a place in any moist border, producing its lemon-yellow globes all through June and July; but it never pleases me so well as when growing wild beside one of our hill burns.

The globe flowers—*Trollius*—are distinguished from *Caltha*—marsh-marigold or king-cup—by the possession of petals. In *Caltha* there are no real petals, the large shining sepals serving for display. In *Trollius* also the sepals are the outstanding feature, the true petals being usually small and inconspicuous, but in the orange *T. Asiaticus* they stand upright and add much to the richness of the blossom.

Foxgloves disagree with lime or chalk as resolutely as do rhododendrons, and those will be disappointed who try to naturalise them on cretaceous soil. But where the soil is free and acid, the difficulty is to keep them in check in gardens wherein they have obtained a footing. The white flowering foxglove is no unworthy companion to the showiest exotics, and there are some beautiful varieties heavily spotted like a

Flowers

gloxinia; but among the seedlings that spring in the borders there are sure to be many of the dull crimson type. These must be ruthlessly pulled out, because, however fine may be the effect of troops of the wild plant in the woods or on rough hill-sides, they give the garden a very weedy appearance. Moreover, insect visitors carry the red pollen to the white flowers. Desiring to ascertain the number of seeds produced on a single stem of fox-glove, I examined a spike of the white variety seven feet five inches high, which had borne 130 flowers. There were about 800 seeds in each capsule, yielding a total of about 104,000 on the spike, a goodly provision for perpetuating the race, but far short of 186,000, the number which Darwin counted on a single stem of *Orchis mascula*.

The common fox glove and its colour varieties are monocarpic—that is, biennial—but the yellow fox-glove—*Digitalis ambigua*—a continental species, is perennial, and natural hybrids between the two kinds appear occasionally in our borders with flowers of a charming shade of flushed apricot. These hybrids are biennial and barren, so it is not possible to preserve the race; but an enterprising florist might secure good results by carefully crossing the two species, a process for which I have neither time nor patience.

There are a few good things in the Onion tribe, and if one keeps clear of ransoms—the wild garlic—*Allium ursinum*—which is a most invasive weed, there are several species that may be safely admitted to the mixed borders. *A. narcissiflorum* (syn. *Pedemontanum*) is the earliest to flower, and shows best when

The Choice of Plants

planted on a bank (we have it on a retaining wall) whence it can hang its fragrant, dull-rose bells. *A. Karataviense* we have lost through mismanagement; it is distinguished for its handsome broad glaucous leaves and large spherical flower heads, grey or lilac in hue, on stems not more than a foot high. *A. acuminatum* has pretty flowers in umbels of various shades of rose, and *A. sub-hirsutum* produces its pure white blossoms in long succession from mid-May to mid-July. *A. Beesianum*, a new species from China, is reported to have fine blue flowers. It is just about to bloom here, and the habit of the plant is most attractive. If it comes up to expectation, I will report in a footnote on the proof sheet.[1]

My favourite in the family—*A. sphærocephalum*—reserves its flowers till August. They are borne in globular heads on slender but stiff stems full three feet high, and are beloved of bumble bees which sleep peacefully on them, rocked by the summer breeze. The grassy leaves die down before the flowers open, leaving the stems with their round heads quaintly like giant hat-pins stuck in the ground.

Howbeit, the loveliest plant in our August borders is *Dierama (Sparaxis) pulcherrimum*. I care not to reckon how many years have passed—they cannot be far short of half-a-century—since I received a root of this from the Edinburgh Botanic Garden; and so generous is the crop of seed that we have never been without it since. As a South African plant, it cannot

[1] It has turned out well, nine inches to a foot high, with flattened, pendent heads of rather light blue flowers.

Flowers

be accounted hardy everywhere; but many of the failures experienced in establishing it are due to attempts to move mature plants. The only sure way is to raise it from seed, lining out the seedlings as soon as they can be handled and planting them out permanently in the second or third year. Provide them with a moist root-run in full sunshine, and they will reward you with six-foot arching wands hung with crimson, pink or white bells that dance delectably in the gentlest breeze.

The seeds of this species are endowed with singular vitality. Sir Frederick Moore having sent me a packet of seed from an unusually high-coloured form grown at Glasnevin, I was so negligent as to allow it to lie on my writing-table for three whole years. It then occurred to me to see if the seeds were still alive; we sowed them in boxes and were surprised when they all germinated at once.

What I conceive to be the wild form of the **globe artichoke**—*Cynara scolymus*—is an outstanding ornament of the border from the time when it unfolds its silvery leaves in May till the stems attain a height of five feet in September and the globes coruscate in florets of loveliest violet-blue. These globes are armed with formidable spines, which cultivation has caused to disappear in the esculent form with which one is so familiar. Our sturdy plant has been in its present position for five-and-twenty years, reminiscent of the time when I served on the Committee of Management of the Chelsea Physic Garden, whence I obtained it, an offset, very likely, from the 17th century original.

The Choice of Plants

The genus *Roscoea*, belonging to the *Scitamineæ* or Ginger order contains some attractive species, interesting also in their peculiar flower-structure. Special care is necessary in their cultivation, for although they are perfectly hardy, and thrive vigorously in free, deep loam, most species are so late in showing above ground that, unless their position is marked by strong metal labels (I find what is called the Shakespeare label the safest) they are pretty sure to suffer from the fork or hoe. *R. purpurea*, from the Himalaya has long been with us and spreads pretty freely. It flowers in July and, like other members of the family, has a peculiar device for securing cross-fertilisation. The anthers lie close up against the hood or upper lip; the stamens supporting them are furnished with projecting spurs, somewhat after the manner of a *Salvia*, against which an insect must press on its way to the nectary. This makes the anthers descend and shed their pollen on the intruder's back and wings. *R. cautlioides*, sulphur yellow, and *R Humeana*, pink, flower in June; followed by *R. capitata*, rich violet-purple, and handsomest of the family so far as known in this country, in August. A plant supplied by Bees under the name of *Roscoea* " August Beauty " flowers here in July and seems to be merely a late-flowering variety of *R. cautlioides*.

The order of Compositæ is so vast and various that I shall not attempt to do more than name a few of those less commonly seen in gardens. The knapweeds are not a popular family, which may be the reason for my never having met *Centaurea Rhapon-*

Flowers

ticum in any garden but my own. I have, indeed, seen a plant grown under that name with pinnatifid leaves; but the true species which I brought back from the Bernina exactly forty-nine years ago and have never been without it since, has entire leaves three feet long, including stalk, bright green above and ivory-white beneath, sharply toothed. It bears rosy purple flowers with a globular involucre of brown bracts, an almost exact counterpart, except in colour, to the yellow *C. macrocephala*. As the two plants flower simultaneously in July, they should be given back-seats in a border near each other.

The Sunflower family tends to coarseness; moreover, some of the perennial species are terrible spreaders, running over a border and exhausting the soil to the detriment of herbs of more refinement; but there is one species—*Helianthus mollis*—which took my fancy five-and-twenty years ago. Mr. Irwin Lynch bestowed a piece of it on me, and we have never been without it since. It grows four feet high and does not ramble, winning favour from all who see it by the contrast between its soft, grey-green leaves and the rich golden flowers which begin to open early in August and continue far into autumn.

Another autumnal-flowering Composite is *Rudbeckia maxima*, which must have received its specific name before *R. laciniata* was discovered, for it has neither the stature nor the coarseness of that plant.[1] From a tuft of fleshy, glaucous radical leaves, *R.*

---

[1] In like manner the American *Rhododendron maximum* was named before anything was known of many loftier species in India and China.

PLATE XI

Natural size   ALSTRŒMERIA HOOKERI   17 August 1918

The Choice of Plants

maxima sends up a stem four or five feet high with a few clasping leaves, bearing one flower, or sometimes two, with yellow rays set round a very dark brownish purple cone an inch and a half high—a very striking herb, which, as a native of Texas, should be given a place in full sunshine and plenty of moisture.

The late H. J. Elwes gave me a delightful little *Alstrœmeria* which he discovered in the pass through which Charles Darwin crossed the Andes, and which he named *A. Hookeri*—a burdensome name for a plant which does not exceed a foot in height. It flowered bravely here in the open in the first year after it was planted, which has enabled me to attempt to give its delicate colouring in Plate XI; but it has never done so in the succeeding three or four years, so I am sorely afraid it does not like us. We would gladly give it a free rein if it were to display the aggressive truculence of others of its kin.

Bœnninghousenia albiflora, or if that name is too thunderous for a human gardener's daily use, let us speak of this nice plant by its synonym—*Ruta albiflora*—is a rue from Nepaul, shrouding itself in a mist of small milk-white flowers in September. It is attractive even before it flowers, for it forms a dome some three feet high, densely clad with finely-divided leaves. It is well worth more attention than it seems to have received, for its appearance is distinct, and one is not too fully furnished with flowers in September.

VII

Some Plants for Walls

EVERYONE who, disregarding John Ruskin's *obiter dictum* that it is only people with minds of the second order who concern themselves with flowers, has given attention to his garden, must be well aware that there are many beautiful plants which, although not hardy in the open border, may pass safely through a hard winter when trained against a wall. That knowledge, however, is not so generally applied in practice as it might be. Often one sees great spaces of wall bare of all covering; in some places walls are overgrown with ivy or that wearisome *Vitis inconstans* which goes commonly by the name of *Ampelopsis*; in other places they are given up to rambler roses and things of inferior merit. Many years ago I received convincing proof of the better purpose to which a wall may be put, even in the colder parts of our country. North Ayrshire—the inland parishes thereof at least—is a cold district exposed to bitter winds, yet in a garden near Kilmarnock, I found the Chilian *Desfontainea spinosa* trained against a wall and flowering freely. Now this brilliant shrub only

Some Plants for Walls

succeeds in the open where the climate is mild and moist.

Nevertheless, one swallow does not make a summer; I must beware of generalising, lest, by recommending things on the border line of hardiness, I mislead those whose gardens are in cold districts. Let the following remarks, therefore, be taken as suggestion for the better use of walls in our milder districts. But what considerable tracts of land fall within that limitation. Kent and Surrey, the whole of Sussex, a great part of Hampshire, all the southern and western coast as far north as Ross-shire, and nearly all Ireland, with favoured patches of climate, such as that of the Moray Firth, here and there on the east coast. I must just jot down a few plants as they come to mind, without any attempt at order, leaving out those which, like jessamine, *Wistaria*, climbing roses, honeysuckle, etc. are too commonly known and grown to require any recommendation.

It is not only against winter cold that a wall serves to protect plants not in the first degree of hardiness; even in an untoward summer it enables them to ripen their young wood more thoroughly than those of the same species in the open, thereby ensuring a fuller flush of blossom in the following year. Very well-marked examples of this may be seen in the present summer of 1923. The Chilean *Tricuspidaria lanceolata* has proved perfectly hardy in the open here; but when first it came to us we were not certain how it might relish a Scottish winter and a harsh spring, wherefore we placed some of the plants against walls,

Flowers

where they have been allowed to remain ever since. The summer of 1922 was so dark and cold that many flowering shrubs failed to ripen their season's growth thoroughly enough to produce much bloom in 1923. Such was the case with *Tricuspidaria* in the open where large plants, sixteen feet high, bore scarcely any flowers, while those with a wall at their back were densely hung with crimson tassels in June. On the other hand, *Abutilon vitifolium*, also from Chile, flowers far more profusely in the open where it has plenty of room, than it does when squeezed up against a wall. *A. megapotamicum* is not hardy here in the open, but does splendidly on a wall.

Those who knew Canon Ellacombe's garden at Bitton will not have forgotten the remarkable variety of plants trained on the long wall standing at right angles to the south front of the house. There grew and flowered the Californian *Penstemon cordifolius*, with brilliant scarlet flowers. It stood, if I remember aright, about eight feet high against a wall facing north. The kind old Canon gave me a plant of it nearly twenty years ago; it has flowered here, but not with anything like the luxuriance it showed at Bitton, probably requiring more sunshine than prevails in our latitude. *Abelia floribunda*, however, a Mexican species, gives us plenty of its crimson tubular flowers against a south wall.

Unluckily for those who dwell in the northerly parts of Great Britain, the brilliant *Tecoma radicans*, from the south-eastern United States, requires a hotter sun than we can provide for it to bring out its

Some Plants for Walls

splendid orange and scarlet trumpets. I, at least, have never seen it in flower freely north of the Trent. It grew vigorously here for many years, clinging to a house-wall by its aerial roots; but, having waited in vain for a display, we removed it to make way for something better suited for our cloudy atmosphere. It is thoroughly hardy, is magnificent when in bloom, and ought to be far more frequently grown than it is in the southern counties of England.

The order of *Leguminosæ* or Pea-flowers contains some splendid plants for training on walls. Of *Clianthus puniceus* mention has been made elsewhere; it has a rival for brilliancy in *Cæsalpinia Gillesii* from Argentina, which runs rapidly over sun-baked walls in districts such as Sussex, the Isle of Wight and even in the home counties, for I first saw it in flower on a south wall in Kew Gardens. Next time I came across it, the effect of its splendour was enhanced by surprise —by the unexpected—which goes so far to intensify sensation. It was in a deserted mining village of Andalusia; a long street of empty houses, mostly roofless, seemed all the more desolate because of the town hall which still stood tenantless, but entire, amid a grove of lofty Australian beef-trees—*Casuarina*—on an eminence commanding the village. No human being was in sight; goats had devoured all edible vegetation within their reach (and for them few green things are inedible); complete silence reigned under the blazing sunshine. Poking about among the ruins I came suddenly, in the backyard of a house, upon a vision of beauty. There, in this dusty, scorched

Flowers

solitude stood a single bush of *Cæsalpinia Gillesii* nine or ten feet high, laden with erect racemes of large, rich, yellow flowers, with scarlet stamens protruding full three inches from the corolla. This sight alone would have been ample reward for a journey to southern Spain; but why should one have to travel so far afield to enjoy it? Why does one never see this noble plant trained against some fraction of the leagues of brick wall given up to common climbers? *C. Japonica* has none of the dazzling splendour of *C. Gillesii*, but its bright green pinnate foliage is very pretty, and when trained on a wall it bears plenty of racemes of deep yellow flowers.

Among the locust trees or false acacias *Robinia hispida* and *neo-Mexicana* are quite hardy in this country, but the branches are so excessively brittle that the plants seldom escape destruction in the open. They make fine wall shrubs with drooping racemes of bright-rose flowers. *Wistaria multijuga* was supposed at first to be no more than a variety of the well-known *W. Chinensis*, to which it is infinitely superior in beauty, bearing racemes of lilac-purple flowers two and three feet long; but it is now recognised as a distinct species, and should always be planted instead of the other, whereof one is apt to get very inferior forms with short, few-flowered racemes.

The perennial species of *Lathyrus* being mostly herbaceous are better adapted for training to a trellis or for covering tree stumps than for growing against a wall, unless they are planted so as to run up among the branches of stronger shrubs. But *L. pubescens*, a

Some Plants for Walls

Chilean species with pale bluish-lilac flowers, is of shrubby habit and well deserves a good place. Far superior to the common everlasting pea—*L. latifolius* and *L. grandiflorus*—is *L. rotundifolius* (syn. *Drummondi*) from Persia, not nearly so often grown as it ought to be, for its flower-clusters of a peculiar shade of soft brick-red never fail to attract attention from visitors. It is quite hardy.

In all but the coldest parts of the country *Solanum crispum*, from Chile, makes a beautiful wall-plant, with a lavish display of bluish-violet and orange flowers at midsummer, exactly resembling those of the field potato. The white-flowering *S. jasminoides*, a Brazilian species, is more beautiful, but more tender, suitable for outdoor walls only in favoured districts. To get this charming climber to mingle its tendrils with *Tropæolum speciosum* so that they may flower together is to achieve the prettiest possible combination.

Like so many other Chilean plants, *Lardizabala biternata* simply revels in our west country humidity. Unluckily the hanging racemes of male flowers are not produced in fair proportion to the handsome evergreen foliage, and as we are short of wall space, and as this rampant climber was not content with covering its allowance of 20 feet by 20, but was invading room allotted to others, I reluctantly signed its death warrant. The male flowers, which are produced in midwinter, form a beautiful object in a vase. Six ivory-white stamens, monadelphously united, stand out well relieved against the broad fleshy sepals and

Flowers

smaller petals, both intensely dark purple, almost black, smelling of vinegar. From the same country comes *Eccremocarpus scaber*, one of the Bignoniaceæ, not rampant, but of free growth, and very gay from midsummer onwards with one-sided racemes of scarlet and yellow flowers. It is easily raised from seed, which ripens in abundance.

Lapageria rosea, another Chilean climber, is not usually accounted hardy, nevertheless there are many places where it flourishes out of doors in this country, and produces its lovely waxy bells in autumn, continuing in flower till Christmas if the season is mild. It should be planted against a north wall, with a guard of perforated zinc round it as a protection against slugs, which are specially partial to the succulent shoots that spring from the roots. The majority of plants that climb by twining their young growth round some support revolve from left to right—against the sun and against watch-hands. Such is the habit of *Convolvulus, Wistaria, Aristolochia, Stephanotis, Berberidopsis*, etc.; but *Lapageria* is one of a few plants, including hop and honeysuckle, whereof the shoots revolve with the sun from right to left.

Berberidopsis corallina, mentioned in the foregoing paragraph, is one of the choicest of the treasures that we owe to Chilean forests. Even if late summer were to pass without its clusters of pendent coral-red flowers, its evergreen, heart-shaped, spiny leaves are so ornamental as to justify a good place for it on a south or west wall.

It is difficult to get away from Chile in discussing

Some Plants for Walls

fine flowers, so rich is that land in desirable species; but I must hurry on. *Fuchsia globosa* and its hybrid *Riccartoni* are well-known as hardy shrubs in maritime districts, requiring no wall to protect them; but there are several species from South America and Mexico of surpassing beauty which I have seen trained as wall-shrubs only in Cornwall and the west, but which ought to be far more generally known and grown. *F. corymbiflora* from Peru and *F. fulgens* are two of the showiest, their long scarlet tubes making them very conspicuous among other things.

Fremontia Californica is noteworthy for its bright yellow flowers set on short stalks along the branches. Though it is hardy in the open in Cornwall and the west, it requires the shelter of a wall in inland districts, and no doubt would be more commonly grown were it not for the difficulty of getting it established through its impatience of being transplanted. I have no experience of this handsome tree beyond having admired it in the gardens of others; but I understand that the surest means of propagating is to raise it from seed and plant it out of a pot. Similar treatment succeeds with *Indigofera Gerardiana*, a deciduous, leguminous shrub from the Himalaya, producing quantities of purplish-rose flowers in racemes during late summer. In cold districts it is nearly always cut to the ground in winter, but springs again and behaves like a herbaceous plant. In one respect it differs from *Fremontia*, which is short-lived and apt to die without apparent cause, whereas *Indigofera* may be relied on to endure indefinitely. Several new species have been brought

Flowers

from China in recent years, some of which are of much beauty.

Abutilon vitifolium having been referred to already (page 78) I need only say in this place that where the climate is too severe for it in the open, no handsomer shrub, and none more profuse in bloom, can be given a good place on a south wall. *A. megapotamicum* is much more tender, but makes a splendid wall plant in mild districts, remaining in bloom longer than any other shrub known to me.

Due advantage is not often taken of a north wall, which is far too often surrendered to ivy or *Vitis inconstans*; but there are many choice plants which rejoice in such a situation, preferring it to any other aspect. *Lapageria* has already been mentioned as one of these; *Camelia reticulata* is another, whereof the great loose, carmine flowers with yellow centres are so much more attractive than the prim florists' varieties which look as if they had been carved in lard, plain or coloured. Rhododendrons of the Maddeni series—*R. Edgeworthi, Scottianum, megacalyx* and others—which it is hopeless to attempt growing in the open except in very mild districts, may be successfully cultivated under the shelter of a north wall, even where winter is sometimes severe. And with what flowers they reward him who bestows this modicum of care upon them! Lily-scented, resembling lilies in shape and purity, the common run of garden flowers seem almost dowdy or aggressively gaudy in comparison with them.

In the *Clematis* family there are plants of the very highest merit for clothing walls. It puzzles visitors

Some Plants for Walls

unversed in botany when they are told that these far-spreading, lofty-climbing plants belong to the same natural order to such diverse herbs, as buttercups, anemones, larkspurs, aconites, columbines, Christmas roses, etc., and therefore are descended from an immediate common ancestor. There is indeed no such polymorphic order of plants as the *Ranunculaceæ*. The beautiful *Clematis indivisa* from New Zealand can only be grown in the open by those whose lot lies in the south and west, where it is not so often seen as it deserves to be, not only for its beauty, but because it flowers in April and May. *C. Armandi*, from Western China, is quite hardy, but demands a vast space to do itself justice. *C. montana* has been in this country for nearly one hundred years, yet it is far from generally grown even now, although as a wall plant or a rambler over growing trees or tree stumps it is hard to beat. The rose-coloured variety, *rubens*, is as good as the type, and equally hardy. It was introduced by Wilson in 1900, and like the white form, flowers in May and June; while another variety, also introduced by Wilson and named after him *Wilsoni*, is described as having larger white flowers, produced in August; but I have no personal acquaintance with it yet, nor with another Chinese species well-spoken of—*C. Fargesi*, a bed of seedlings thereof here not having reached the flowering stage. *C. flammula* has probably been cultivated in this country longer than any other exotic species, having been brought hither during or before Queen Elizabeth's reign; and although the flowers are individually small they are produced in great

Flowers

abundance in autumn and diffuse a delicious fragrance.

Among the large flowering species—*C. patens, lanuginosa, florida, Jackmanni* and their innumerable hybrids there is plenty of choice, not for clothing walls, for these are not of the far-flung character of the kinds mentioned above, but for decorating them. They may be planted to climb among the branches of other shrubs, whether these are growing in the open or trained against walls. But they must not be starved; a good mulch of well-rotted manure is no more than their due; they all relish lime in the soil and full exposure to sunshine, provided that their roots are kept in shade. I am not competent to prescribe for the niceties of pruning appropriate for the different species, that process having been unpardonably neglected here; so I cannot do better than refer the reader to the admirable treatise by Mr. W. J. Bean, where directions founded upon experience at Kew are well set forth.[1] In the course of the present summer, nothing in our garden has pleased me better than the pure white flowers, six inches across, of a clematis of the *Lanuginosa* race, rambling to a height of twelve feet over a myrtle trained on a south wall. It has rejoiced us in the same gracious manner during the last fifteen summers, has never been pruned, and we dare not give it the lime which it is known to covet lest that should disagree with *Amaryllis belladonna* at the foot of the wall. Those who love *Clematis—*

[1] *Trees and Shrubs hardy in the British Isles*, vol. i. pp. 351-369 (first edition).

Some Plants for Walls

and who that knows the family can fail to love them?
—should study for guidance the enthusiastic monograph by Mr. William Robinson, entitled *The Virgin's Bower*.[1] And whereas the purpose of the present writer is to assist in rejection as the indispensable corollary of selection, let his last words about *Clematis* be to denounce all double-flowered forms of that family as abominable. No milder epithet suffices for the distortion of a perfect flower into a monstrosity that can only merit the kind of morbid attention bestowed upon a two-headed calf or a four-legged chicken in a village museum.

The passion flower—*Passiflora cœrulea*—is a rapid grower and, although a native of Brazil, is quite hardy as a wall plant in all but the colder districts; but in the northern parts of our country it flowers too late in the season to form its egg-shaped orange fruit which are so ornamental on house-fronts in some of our southern seaside towns. The white-flowering variety, "Constance Elliot," is of inferior beauty to the blue-flowering type.

Lonicera tragophylla, introduced by Wilson from China in 1900, is distinguished among other honeysuckles by the size of its flowers, which are borne in heads of ten to twenty blossoms of a uniform rich yellow averaging three inches long. It differs also from most other climbing species in being scentless; nevertheless, it is such a handsome plant when in flower as to make it worthy of a place in the most select assemblage of plants. But let no one now make such a

[1] London, John Murray, 1912.

Flowers

blunder as I was guilty of when first it came here. It was planted against a south wall, where it has become a prey to a dark-coloured aphis; the proper place for it being a north wall, on a cool soil.[1] Unluckily, we cannot now remove it, as a precious plant of *Mutisia decurrens* has wound its tendrils inextricably among its sprays. *L. Standishi* and *fragrantissima* are the reverse of showy and resemble each other very closely; in fact, the only ready means of distinguishing between them is that *L. Standishi* has whitish bristles on the stalks of leaves and flowers. They both bear slender, white, very fragrant flowers in long succession through the winter months and are worth growing on that account alone.

Carpenteria Californica yields its large, white, yellow-centred flowers more freely against a wall than it does in the open; and if a tangle of *Tropæolum speciosum* gets among its branches, you will have one of the most charming accidental effects possible, deliciously suggesting strawberries and cream.

Ceanothus rigidus and *Veitchianus* with rich blue flowers and *C. thyrsiflorus* with pale blue ones, are desirable evergreens for covering walls, though the last-named is hardy enough to stand out in the open even in cold districts, and grows over twenty feet high. *Ribes speciosus* is deciduous, but the brilliancy of its flowers entitle it to a choice corner; while *Abelia floribunda* can only be trusted to weave its long coral neck-

---

[1] That, at least, has been our experience here, and that of others elsewhere; but at Tresco it flourishes against the south face of a rock in blazing sunshine.

Some Plants for Walls

laces in such situations as the lemon-scented verbena—*Lippia* (*Aloysia*) *citriodora*—can bear it company, they being about equal in hardihood and essentially plants for the south and west.

I will wind up this disorderly chapter with a recommendation of the evergreen *Magnolia grandiflora* as noble tapestry for the south wall of a spacious mansion, where it may fill the air with perfume from its great cream-coloured chalices. Unhappily for us northerners, it flowers too late in the season for our latitude; at least I have never seen it flower in Scotland; but in southern counties it is without a peer among flowering trees.

It may be observed that no mention has been made of roses as wall plants. That is assuredly not from any indifference on the part of the writer to that race of plants which, more than any other, has been the joy and pride of countless generations, but it is because to deal with them, however superficially, would lead me far beyond the purpose I have expressed in the fore-note to this volume. Nobody will peruse these pages who takes no interest in gardening; while those who are interested in the subject will have their own taste and preference in roses, and have unlimited choice of varieties to choose from. Nevertheless, I will break the rule I have laid down for myself by naming one rose, not as the only one best fitted for a wall, but as one that has afforded us special pleasure and does not seem to be very generally grown. The rose in question is the hybrid tea Climbing Papa Gontier. Its carmine flowers may or may not be of the regulation pattern prescribed by experts; of that

Flowers

I am quite ignorant; but this I know, that it has grown some twelve feet high beside our dining-room window, bidding us morning greeting earlier than any other rose in the garden, while winter still lingers in the lap of May, and in chill October waves reluctant farewell to summer joys.

VIII

Rockwork and Edgings

I APPROACH the subject of rock-gardens with the trepidation that becomes one who finds himself out of touch with the times. Successive generations of landscape gardeners having exerted themselves to divest our pleasure-grounds of every natural asperity, turning heathland into closely-shaven lawn, quarrying away every outcrop of rock and draining off everything like bog, the abandonment of bedding-out was followed by a vehement naturalistic reaction, and now no garden is reckoned up-to-date that is not furnished with a rockery. Mimic Matterhorns have been piled in sylvan recesses of the Thames valley; ravines have been carved out of gentle slopes in the richest wheat-lands; miles of piping have been made to induce water to course through make-believe cañons, and thousands of tons of rock have been imported from Yorkshire and Westmorland to rear miniature Andes in Berks and Bucks.

I was taken one day to inspect a huge rockery to which a gang of workmen were putting the finishing

Flowers

touches. It was in the grounds of a splendid mansion that had been recently erected where

> The great black hills, like sleeping kings,
> Lie grand round Rothesay Bay.

An amphitheatre of Highland mountains caused this laborious make-believe to appear the reverse of impressive; but I congratulated the foreman upon the skill with which water had been led from a considerable distance to form a rivulet in this diminutive wilderness. "Ah," said he, "it *is* wonderful what our firm can do with a three-inch pipe!"

It is a pleasing pastime, no doubt, this mimicry of the grandeur of nature; so is the tending of dolls' houses and Noah's arks for heads and hands at a different stage of existence. It marks how sentiment for savage landscape has changed since an eighteenth-century bard, returning from a tour in the Highlands of Scotland, thus frankly expressed his feelings:

> Bleak mountains and desolate rocks
> Were the wretched reward of our pains;
> The swains greater brutes than their flocks,
> And the nymphs as polite as their swains.

Those who possess within their grounds features such as the wonderful quarry in Mr. F. J. Hanbury's garden at Brockhurst or the limestone cliff in the late Mr. Farrer's nursery at Clapham, in Yorkshire, would miss a stroke of good fortune if they failed to avail themselves of a rockery ready-made. But such examples are rare; in most cases artifice is but too

Rockwork and Edgings

obvious in what is designed to represent savage scenery. To simulate

> The grisly rocks that guard
> The infant streams of Highand Dee

amid the heavy-headed elms of Warwickshire is to court discomfiture such as Robert and James Adam incurred when, abandoning the dignified Georgian architecture wherein they excelled, they pranked out quiet country houses into meaningless embattlements and innocuous *machicoulis*. If gardening is to rank as an art it must be subject to the law that bans all sham and make-believe.

Let me not be understood as under-estimating either the beauty of alpine flowers or the cultural value of rocks and stones in the soil. My feeling is that alpine plants are best grown without puny attempts to emulate alpine scenery, and that plenty of stones contribute to the welfare of all kinds of plants in the herbaceous border. The majority of alpine plants are satisfied when provided with a deep root-run, moisture during the growing season, rest as complete as may be had in winter and effective drainage at all times. In the plan that we have adopted at Monreith with moderate success all these requirements have been met so far as possible in such a variable climate as ours. Flanking a long terrace facing south and west were steep grassy banks troublesome to mow and devoid of interest when mown. Owing to the configuration of the terrace, sections of these banks were turned to various points of the compass. Finding it desirable to get rid of the scythe work necessary to keep them,

Flowers

we stripped off the turf, faced it with rough stones as a retaining wall, and planted it with alpines and other suitable things. That was about thirty years ago, and it really has been a source of never-failing pleasure ever since. Of course many treasures have perished, others have been suffocated by ruder neighbours, and constant attention has been necessary to expel dandelions and other weeds; but on the whole it has met all the requirements of a regular rockery. The plants root into the soil behind the stones which protect them from summer drought, and the steep slope ensures thorough drainage in wet winters. Still there is something wanting to make a perfect arrangement. A perforated pipe ought to be laid from the water tank along the top of the banks to discharge a gentle deluge among the roots in dry weather, simulating the action of melting snow. This would have been done long ago, for the tank is in position, fed with water from the present writer's bath; but to lay the pipe it would be necessary to uproot many precious shrubs and other plants that grow in the borders above the wall, and we have not mustered the necessary hardihood for such an upheaval. Banks of the kind described are of frequent occurrence in many gardens; saving of labour is a necessity in nearly all, and to that end the operation above described tends very materially.

A few of the plants may be named that have contributed most to the attraction of our rock wall. *Aubretia*, of course, avoiding the varieties that incline to red, leaving that note to be struck by *Phlox subulata*

Rockwork and Edgings

and *verna*, and using *A. Græca* for lavender and "Mrs. Lloyd Edwards" or "Dr. Mules" for purple. *Aubretia* must be watched lest it over-run rarer things. So must *Alyssum saxatile*, a common herb, but indispensable for contrast with *Aubretia*. *Arenaria montana* best of the sandworts, with its pure white stars; *Dryas octopetala*, whereof *D. lanata* is a variety more prodigal of flower, enjoys some lime, while *Lithospermum prostratum*, with its lovely variety, "Heavenly Blue," detests it. So, probably, does the June-flowering *L. graminifolium*, but it has not been given a chance here of expressing its opinion upon that condiment.

Among the St. John's-worts, the best for our work is *Hypericum reptans* (not to be confused with the far inferior *H. repens*), which hangs as a wide mat of neat dark foliage, opening its golden eyes in July and scarcely closing them before the winter gloom. *H. fragile* is more showy when in bloom, but its flowering season is shorter. *Spiræa decumbens* is the only one of that family for our present purpose, but it makes a fine curtain on a rock wall facing west or north, decked with rosettes of white flower throughout July. The rock roses—*Helianthimum*—are indispensable for their mid-summer glory of gold and white, and almost every shade of red except scarlet; but they require to be watched, so freely do they scatter seed and invade crannies assigned to rarities.

So much for the general drapery of the rock wall. It must be kept in control lest it over-run the spaces reserved for less exuberant growths. The difficulty is not to find furniture for such spaces, but to make

Flowers

choice out of the abundant variety at our disposal. A whole chapter might be filled with discourse about Saxifragas, from the diminutive *S. Irvingi* with shell-pink flowers in February, to the stately *S. longifolia* with its massive plume like the brush of an arctic fox and the looser spray of *S. cotyledon* flung out at midsummer. *Dianthus callizonus* prefers a westerly or northerly aspect, but *D. neglectus, alpinus, squarrosus* and the maiden-pink—*D. deltoides*. The same may be prescribed for the cheerful little alpine catchfly—*Silene alpestris*. Then there are the houseleeks—*Sempervivum*—and the stonecrops—*Sedum*—little people so patient of drought that they may be popped into any vacant space at the top of the wall; while *Ramondia Pyrenaica* shrinks from sunshine, but lights up a dark moist recess with delicate purple posies. A north aspect or, failing that, a westerly one, is best for those two azure beauties *Meconopsis latifolia* and *simplicifolia*[1] (Plate V). These do quite as well in the open border as anywhere, but never show to such advantage as when perched on a moist wall, and happy is he who has got them naturalised in such a position. They are monocarpic as a rule, that is, they die after flowering once, but sow themselves freely in places to their liking and, self-sown, they produce far finer flowers than those raised in boxes to be planted out; for, like most of the Poppy order, these Asiatic beauties intensely dislike having their roots disturbed.

[1] The ordinary type of *M. simplicifolia* has dark blue flowers, sometimes tending to purple, but in the variety known as Baillie's they are of the most exquisite sky-blue.

Rockwork and Edgings

Two herbs of humbler proportions may be trusted to establish themselves without interfering with other plants. *Erinus alpinus* never makes itself more thoroughly at home than on a rock wall such as we have been discussing. The typical form has flowers of soft rosy-mauve, the variety *carmineus* being clear carmine. Both are desirable; it is hard to decide which is the better; but it is well to give each a range of wall to itself, for the effect of the flush they produce in June is marred by a mixture of the tints. There is a white flowering form which, like a white-flowering forget-me-not, ought to be scrupulously weeded out. All that is necessary to secure this plant in permanence is to insert two or three seedlings among the stones, leaving them to attend to their own propagation, which they do by scattering their seeds along the face of the wall. In two or three years they will have spread all over it. It is easy to pull it out when it encroaches upon plants of less vitality, for the roots take but a moderately firm hold and the foliage is the reverse of dense.

Equally innocuous is the little annual *Saxifraga Sibthorpi* (syn. *cymbalaria*), a native of Greece, which is not so greedy of sunshine or so patient of drought as *Erinus*, but will quickly colonise a wall facing north or west, and lighten it up in spring with innumerable yellow flowers.

Campanula pusilla may be trusted to run through the chinks of the moister parts of a retaining wall that has a good root-run behind it. One of its synonyms is *C. modesta*—most appropriate to its delicate heart-shaped leaves, which are apt to escape notice until in

Flowers

July crowds of fairy bells are hung out on four-inch stems, dark blue, light blue or white according to the variety. It is so hard to say which is the prettiest form that the problem should be solved by growing all three. There are many other species of bell-flower suitable for a rock wall; it would lead me too far to enumerate them; better service cannot be rendered to the reader than to refer him to Farrer's *My Rock Garden*, wherein Chapter X is allotted to describing and prescribing for " the Smaller Campanulas."

Moltkia petræa is a bulkier plant, a Dalmatian subject rejoicing in full sunshine, well deserving a ledge pretty high up on the wall for the better display of its blossom. Formerly classed among the gromwells—*Lithospermum*—it appears in the Kew *Hand List* as a herbaceous plant; but Mr. Bean, Curator of Kew Gardens, gives it honourable mention in his admirable work on trees and shrubs, company to which, no doubt, it is well entitled, for although its full height may be reckoned at eighteen inches, it has a stout woody stem and spreads to double that in width. It is a lovely object when crowded in July with clusters of blue flowers.

Now I must not be drawn into a treatise upon alpine plants, a subject which has been thoroughly dealt with by abler hands. My purpose is no higher than to indicate methods of growing them which does not entail the heavy labour required for the construction of an orthodox rock garden, yet enables one to enjoy the varied charms of this class of plant with the utmost ease.

Rockwork and Edgings

Let me turn to another system under which many of the choicest alpines may be grown with excellent effect. Herbaceous borders may be laid out in either or both of two ways, namely, with grass lawn reaching up to and along their margin or with a gravelled or paved path alongside of them. When the latter is the arrangement, an edging of some sort is indispensable, and in the material of that edging there is opportunity for a great variety of treatment. Box is of immemorial use for the purpose, and the patience with which it submits to the shears renders it unequalled for outlining a formal design. But whereas many of us prefer informality in a flower pleasaunce, and whereas it is greatly to be desired that uniformity in design and furniture of gardens should be avoided, there is hardly any feature offering more scope for difference than the edging of borders. Example being better than precept, the simplest way to explain my meaning is to describe how we have been proceeding in this matter.

Have been, I say; forasmuch as the process is one to be accomplished by degrees. Having several borders separated from long stretches of gravelled path by a two-foot margin of grass, troublesome to mow, we began by stripping off the turf, which was carefully stacked to form loam. The soil underneath was well loosened and fresh loam laid over it, whereon were laid rough stones weighing 20 lb. to 30 lb. each, leaving just enough space between them to receive the plants intended ultimately to grow round and over them. Both sides of a broad terrace-walk twenty-

Flowers

four yards long were treated thus in the first year, one side being planted with mossy saxifrage, red, pink, and white; the other with gentianella—*Gentiana acaulis*. Very gaunt and unpromising seemed the edgings till the plants began to grow; but now the saxifrage has hidden all but the moss and lichen that clothe the tops of the stones, becomes a fine band of colour in April and May, and most agreeable verdure at other seasons. On the opposite side the stones remain in evidence, the gentianella growing round, but not over them, and forming a brilliant belt of blue in April, with a few scattered blooms throughout a mild winter. So satisfactory was the effect, that a fresh length of edging has been treated each year in a similar way, using a variety of different plants. Among those which are most effective may be noted encrusted saxifrages, especially *S. aizoon rosea*, *Campanula Portenschlagiana* (ten syllables to denote a plant six inches high!), *Æthionema* Miss Willmott, *Thymus serpyllum micans* and *coccineum*, *Saxifraga apiculata* and *Andrewsii* and *Dianthus neglectus*, the last-named being in constant peril of being weeded out, so closely does its foliage resemble some kind of grass. The edging may be enriched by planting dwarf bulbous things close behind it, either in a continuous row or dropped in here and there. The other plants will cover them, but such things as spring and autumn flowering crocus, winter aconite, *Scilla Siberica* and *bifolia* will push up through them and tactfully wither out of sight when they have played their part. Pansies and violas make very

Rockwork and Edgings

rich edgings by themselves, but must not be planted along with the alpines, which are entitled to full exposure. The brightest varieties of *Erica carnea* are good edging plants without requiring or being helped by the use of stones. Canon Ellacombe convinced me of the advantage of clipping this heath after the flower is past, and this entitles it to an occasional topdressing of peat, sand and loam shaken well in among the stems.

Several species of stonecrop make very neat edging and flower profusely. *S. spurium*, beloved of bees, remains longer in flower than most of them, but none should be grown except the variety with rich crimson flowers, avoiding that in which they are of a washy pink tint. Our native *S. Anglicum*[1] and *acre*, the first with pink-tinted, the second with yellow flowers make a brave show when in blossom; but are too scantily furnished with leaves to be recommended for edgings. They should be encouraged as weeds in the rock wall. A species which arrived here (whence I have forgotten) bearing the name of *S. Guatamalense* has flowers very similar to those of *S. Anglicum* and better foliage.

Let him who would avoid trouble avoid another of our native stonecrops—to wit, *Sedum rupestre*, which is actually recommended in a book on garden flowers now lying before me as " a good edging plant for borders." A useful book, too, on the general

---

[1] The name *Anglicum* is not very appropriate, for this pretty plant is not common in England, but extremely so on the west coast of Scotland and Wales, and in Ireland.

Flowers

subject, but let me nail this misstatement to the counter at once. In gardens this stonecrop is a pestilent weed; woe worth the day when it obtained admission to ours, for now a lot of precious time is squandered in trying to get rid of it. You may uproot the wretch, as you think successfully, but it is sure to drop some of its linear, succulent leaves, each one of which will strike root and come up smiling as a vigorous stripling. Fancy the synonym for this coarse member of an estimable clan being *Sedum elegans*!

Prunella Webbiana, under which name we received a plant from Wisley, I presume to be only a variety of *P. grandiflora*; if so, it is a notable improvement on the type. The profusion of its rich deep violet blossoms, lasting a long time, makes it an admirable edging plant, and if the flower-spikes are removed as they fade, a second bloom is thrown up in autumn.

China has sent us a species of celandine—*Chelidonium Franchetianum*—which makes an edging as curious as it is beautiful. The leaves, pinnately cut into broad lobes, are of the colour known to painters as *terre-verte*, blotched with pearly grey. They lie flat along the soil, and have the faculty of retaining through the hottest summer day the dew or rain that collects in beads upon them. If it is desired to keep the edging specially trim, the plants may be prevented from producing their yellow flowers in May or June. *Silene Schafta*, from the Caucasus, makes a beautiful band of bright rose-colour in late summer, and is well suited for edging a border. It requires time to get established before doing itself justice.

IX

Some Failures

On apprend en faillant.—*French Proverb.*

IT is very meet and right to chortle over success with choice plants; we are (at least *I* am) ready enough to do so, all the more so, perhaps (for such is the frailty of human nature) if we hear of neighbours and friends failing with them; but it is equally profitable, if not our bounden duty, to chronicle failures; wherefore let me—for once—make a clean breast of it.

We are all tempted to try and grow things which are a little short of being hardy in our climate, and must not be surprised if the percentage of loss in such plants is high. But there are certain plants which although capable of enduring any amount of cold that they are likely to encounter in the British Isles, obstinately refuse to flourish in some gardens, although they are quite happy in others, sometimes only a few miles distant.

For instance—how comes it that for a full quarter of a century I have been employing every available device to grow the common *Anemone hepatica* without

Flowers

a gleam of success; whereas in a neighbour's garden, six miles away, on the same Lower Silurian formation and in similar proximity to the sea, it forms great clumps and carries abundant leafage through the summer. Many years ago the young ladies who then lived there removed quantities of hepatica from the borders and planted them out in the woods, where they were promptly devoured by rabbits. Again—I have in mind a garden near Jedburgh where the common blue hepatica not only grew in crowds within the garden, but had strayed beyond it and had populated a steep bank overhanging the river. In that place the rock is old red sandstone.

Another plant of this genus which has persistently responded to the most sedulous coaxing by a blunt negative is the bright *Anemone palmata*. We have given it all the sunshine that is to be had in our sloppy climate, but all in vain. A few flowers appear the first season after planting; just enough to make one wish and hope for more; but thereafter there is no sign of life save scattered half-hearted leaves. In this case, however, there is no obscurity about the cause of our failure. That was revealed to me in May, 1921, in a motor drive from Lisbon to Badajos. Between Setubal and Casa Branca the road—or what the freakish imagination of Portuguese local government causes them to designate a road—passes over an elevated moorland plateau, whereon the beauty and variety of wild flowers lured us into stopping the car several times. One of these halts was made at a place more bare and sun-baked than the rest, where a space

Some Failures

of two or three acres was thickly set with *Anemone palmata*. The foliage was nearly all withered; only here and there remained a few blossoms, both yellow and white; all the rest had gone to fluffy seed. Three weeks earlier there must have been a fine display. Here was revealed the secret of our want of success with this wind flower. We cannot—on the west coast at least—secure for it torrid sunshine and a heavy soil baked to the consistency of a pantile. To the same climatic cause must be attributed the positive refusal of any species of *Sternbergia* even to exist, still less to flower, with us. "All species," says Mr. Weathers in his excellent *Bulb Book*, " are perfectly hardy, and will flourish in any good garden soil . . . When grown in bold masses in the rock-garden, shrubbery, grass-land or flower-border, they are wonderfully effective." That is the sort of seductive description that has beguiled me into reiterated experiment with this bright autumn flower, all to no good. The conditions in Surrey, where I have gloated over its golden glitter, are very different from those of the humid west, and I have never seen it in better condition than on the sunburnt selvage of a vineyard in the Médoc.

There are several herbs which respond freely enough to more vapoury sunshine than prevails over their native hills and plains, but which cannot survive the drenching winters of the west. They may be brought safely through the dark months by potting them up and keeping them in a cold frame; but that is altogether at variance with the permanent character of a

Flowers

herbaceous border or rock-garden. There is such abundance of good things that flourish under our dripping skies that we can afford to do without those that will not do so. Wherefore we have written off as irreconcilables the delectable Golden Drop— *Onosma Tauricum* and its near relative, *O. albo-roseum, Penstemon confertum* and *heterophyllus, Myosotis rupicola, Linum capitatum, Silene Elizabethæ* and *laciniata Purpusi, Malvastrum coccineum,* the blazing magenta *Calandrinia umbellata* and the lovelier *C. grandiflora.* Over these we shall worry no more, believing that nothing will ever reconcile them to our winter wet. But why should *Brodiæa volubilis* disappear after delighting us with its long, twining cord-like stems with dangling umbels of clear rose? Farrer's Chinese *Cyananthus* No. 1220 has proved to be more patient of wet when at rest than the Himalayan *C. lobatus* which we have long since dismissed as hopeless.[1]

> Green's forsaken and yellow's forsworn,
> But blue's the prettiest colour that's worn.

So runs the old distich, and one parts with a blue flowering herb more reluctantly than with those of other hues. I have never been so daring as to attempt *Eritrichium nanum*—that will-o'-the-wisp of the Alps —but times and times have I applied every device to persuade *Gentiana verna* to abide with us, always with the same dwindling result. After all, disappointment

---

[1] I am told by a fellow-amateur in Argyll, where the climate is very similar to ours, that *C. lobatus* grows luxuriantly with him, and that I must have erred in not giving it deep and stony soil.

Some Failures

with that piece of living lapis-lazuli is amply balanced by the profusion with which the gentianella—*Gentiana acaulis*—responds to March sunshine with its thousand azure liqueur glasses. One hears complaint sometimes that in certain gardens this miracle of splendour does not prosper. The only care it gets in my poor kaleyard is uprooting, division and replanting every four or five years after it has formed such a dense mat as to interfere with the production of flower.

If the late Mr. Reginald Farrer had sent home no other spoil from his expeditions in China than the gentian which bears his name, he would have earned the enduring gratitude of stay-at-home amateurs, for *Gentiana Farreri* is the loveliest of all the autumn flowering species of that genus. It is a glorified *G. ornata*, excelling that kind in vigour (in some gardens, at least) and the size of its flowers, which are sky-blue, of the same celestial quality that distinguishes Captain Baillie's variety of *Meconopsis simplicifolia* above all others of the poppy tribe. The interior of the throat is pure as the white of an egg, the outside palest buff, vividly striped with dark blue—almost black—lines, like the waistcoats worn in undress livery by footmen before they all went off to the trenches in khaki. If I hesitate before recording failure with this plant I cannot honestly register success, for although it has flowered here sparingly in five successive autumns, several plants have gone to their winter sleep and failed to awake in spring, and the survivors have not as yet shown any tendency towards spreading into such a

Flowers

mat as inflames one's envy in the rock-garden of the Edinburgh Botanic. *Spero meliora!*

It was after many failures that we at last got another blue flower established, viz., *Mertensia Virginica*. It is worth all the pains bestowed upon it, for it is the queen of the genus. *M. Siberica* grows and spreads amain, and would prove a worthy rival of the other if its sky-blue flowers were as large as those of its darker American cousin, which always used to become conspicuous in the second spring by its absence behind the label, till we met its fancy by adding a little lime to the loam and peat provided for it. Evidently it cannot put up with an acid soil. *M. echiodes, elongata* and *lanceolata* must be added to our lengthening list of lost ones, but in their case we have our own negligence to blame, and I console myself for their departure by gloating over the glorious blue of *Cynoglossum nervosum*, which requires no attention —only to be left alone. *C. appeninum* is even better, flowering a month later, *i.e.* in June, and giving a second blow in autumn if the first lot are cut off when fading. Yet I read in a work on horticulture of much merit that the genus *Cynoglossum* consists " of rather coarse, tall-growing biennials or perennials, scarcely suitable for the flower garden; " which reminds one how it was said of old that if there were no difference of opinion there would be no fancy waistcoats.

The Composites as a rule are a friendly clan, some of them, like the daisy, the dandelion and the thistle resisting all measures to keep them at arm's length; indeed, the only member of that order which

Some Failures

declines to take up its abode with us is the purple cone-flower—*Echinacea purpurea*. Sometimes it has endured for two or three years, obviously under protest against something in soil or climate unacceptable for a native of Louisiana; bearing aloft just as many of its quaint, sober-toned blossoms as cause one to neglect no effort to please it; but never yet has it put in an appearance in the fourth year. I am seldom in southern England in late summer or autumn when this species is at its best, which may be the reason for my never having seen it thriving as a Composite should do. I would accept gratefully sound advice from anyone whose experience has been happier than mine, for the ray florets of this *Echinacea*, set round the long, swarthy cone, are of a shade between *vieux rose* and brick-red which it is not easy to match.

Mutisia decurrens, that most eccentric of Composites, has just narrowly escaped being reckoned in our black list, for although we have found it an extremely difficult subject to establish, one plant has rewarded us late in every summer since 1916 with its great orange blossoms, reminding one of the rising sun on the national flag of Japan. It is now nine feet high, thriving vigorously, pushing its steel-grey leaves and tendrils through *Cotoneaster horizontalis*, which gives it the support it requires. The genus *Mutisia*, besides its beauty, is of much botanical interest, being the only member of the Compositæ that climbs by leaf-tendrils. More than thirty species of the genus are known; *M. decurrens* having been introduced to this country by Messrs. Veitch between sixty and

Flowers

seventy years ago. A native of the Chilian Andes, it might be expected to respond to cultivation as readily as do so many plants from that region. That it is not impatient of cold may be judged from the way it has flourished for many years in the Edinburgh Botanic Garden, where the winter climate cannot be described as mild. Probably the only reason for this fine climber not having found more favour with amateurs consists in the difficulty of getting it to start after it is planted out from the pot. We have lost at least half-a-dozen at that critical time. It was some time before the secret of propagating it was discovered. September rains always rot the seed at Monreith before it ripens, although in a cool house in the Edinburgh Botanic Garden it produces fine fluffy heads of seed, which seems all right; and in the *Notes* of that institution for November, 1913, Mr. Laurence Stewart describes the success he has achieved in taking cuttings after the flowers have faded in early autumn. This is valuable information, forasmuch as *Mutisia* intensely resents mutilation at other seasons. Let no one be tempted to lift the suckers which it produces in free loam and peat, for such treatment is apt to bring about the death of the whole plant.

The blossoms of *M. decurrens* are very showy and of peculiar construction. They measure about five inches across, and consist of eighteen or twenty ray florets of soft, but rich, orange, reflexed at half their length, with a yellow disc in the centre round which a circle of dark maroon florets rise like a palisade. There is a coloured plate, as beautiful as it is faithful,

Some Failures

of this species in the *Garden* for 22nd December, 1883. The only other species of *Mutisia* which we grow here is *M. ilicifolia*, with washy pink flowers of very inferior beauty to *M. decurrens*, but the grey foliage, shaped like small holly leaves, is attractive.[1]

But I must hark back to our failures, among which must be numbered the attractive little *Potentilla Tonguei*, a garden hybrid of origin unknown to me. It is described on good authority as " of very easy culture in sunny places," but hitherto all that it has vouchsafed to us consists of just enough of its salmony-orange blossoms blotched with maroon to make us wish for more, before wilting and disappearing. Then there is that brilliant little Californian willow-herb—*Epilobium obcordatum*—which Reginald Farrer so aptly described as " a good plant in a family filled with iniquity." Over and over again have we planted it, subject to the conditions prescribed by those learned in alpines It gives us in the first season a few of its cherry-pink flowers, so large in proportion to its own dimensions, but in the second year the place thereof knoweth it no more. Another species, pretty but not nearly so fascinating as the Californian beauty, behaves nicely. It came to us under the name of *Epilobium Fleischeri*, which the Kew *Hand List* enters as a discarded synonym of *E. Dodonaei*; but on turning up the plant so named in horticultural works I find it described in one of them as growing two and a half to

---

[1] Since writing the above I have seen at Truro flower show blooms of *M. clematis* from the Rev. A. Boscawen's garden at Ludgvan, in the favoured neighbourhood of Penzance.

Flowers

three feet high, and in another as one foot high. Our plant cannot aspire to more than nine inches, so there seems to be some confusion among experts.

With rhododendrons we have had remarkably few disappointments among about one hundred and fifty species growing here, besides hybrids without number. *R. Fortunei* var. *cyanocarpum*, one of Forrest's later introductions, has failed unaccountably; two plants, having thriven vigorously to a height of three feet, suddenly sickened at midsummer; the beautiful myrtle green leaves, with glaucous backs turned yellow and dropped off. *R. Kamschaticum* is a dwarf deciduous shrub with a vast geographical range, extending from British Columbia, through Alaska, along both shores of the Behring Straits and as far south as north Japan. It is no novelty, having been brought to this country in 1799; but the only place I have seen it flowering—and there profusely—is in the rock-garden of Edinburgh Botanic Garden. The late Sir Isaac Balfour gave me a good piece of it, and it has passed through eight or nine seasons here, apparently in the best of health, but has not yielded one of its crimson blooms which, measuring as they do $1\frac{1}{2}$ to $1\frac{3}{4}$ inches across, are preposterously out of proportion to its stature of six or eight inches. Imagine what a sight *R. arboreum*, thirty feet high, would present if it bore flowers on a similarly proportionate scale!

Has any one achieved success with *Stenanthium robustum*? If so, will that fortunate individual communicate the secret through the horticultural press. In its first season here it justified its title

Some Failures

robustus by throwing up plumes two feet high, reminding one of the *panache blanc* of Henry IV. or Stilicho's *cognita canities*; but it never renewed the display; nor have subsequent attempts to establish it confirmed our expectation of what promised to be a notable acquisition for the autumn border. It belongs to the Lily order, and I fancy must be of Asiatic origin, for in 1908 Reginald Farrer wrote about it as a novelty— " so new that I can say little about it "—and that enthusiast was never at a loss for words or will to convey information.

Most plants of the order *Campanulaceæ* contain some essence peculiarly attractive to slugs, and it is to the gluttony of that clandestine race, not to any fastidiousness on the part of *Phyteuma comosum*, that we owe our inability to keep that fantastic form of bellflower. The field slug—*Limax agrestis*—light brown with dark longitudinal stripes—is its arch-enemy, and will clear off a whole clump in a short summer night. *Adenophora megalantha* is another desirable plant of the same natural order, recently brought from the far east, and of far superior beauty to any other of its genus. It bears pendent bells of a good blue, twice the size of the common harebell, supported on stiff fifteen-inch stems that never flinch or break in any weather. It quite won our hearts by flowering generously in July of two or three consecutive years, until from some secret ailment it disappeared, leaving me to reproach myself for having neglected to save seed. It came here from Bees Limited, a firm which in its early years was ready to supply many novelties

Flowers

of merit, but which seems to have received so little encouragement from the meagre demand for anything out of the common that it has lapsed into the jog-trot of ordinary nursery stuff.

Acantholimum venustum, choicest of the prickly thrifts, did well for a while with us on a retaining wall, which is all we have in the way of rockery (and far from unsatisfactory at that), but became disgusted with our unreasonably wet winters and left the field to the less ornate, though still commendable, *A. glumaceum*.

Bitter in proportion to the magnificence of *Embothrium coccineum* has been the disappointment attending all our attempts to grow it here. Promising young plants have been put out in choice places on several occasions, and have refused to take on. It is admittedly an uncertain subject to establish; but it flourishes in three gardens within our neighbourhood, which renders our failure the more grievous to be borne.

I have kept the lilies to the last in this melancholy confession of misadventure, because in no genus of plant is the proportion of ill success so high. It varies with soil and climate in respect to different species, but the outward and visible sign of distress is everywhere the same, to wit—the appearance and spread of the fungus *Botrytis cinerea*. It is very commonly assumed that this fungus is the *cause* of disease, and sometimes of death, in lilies, whereas in fact it does no more than take advantage of the vitality of the plant having been lowered to a degree rendering it a suitable

Some Failures

nidus for the parasite. A parallel instance of misinterpreting pathological phenomena prevailed for very many years in relation to the salmon disease. Even so keen-sighted and practised an observer as Thomas Huxley confidently pronounced the fungus *Saprolegnia ferax* to be the sole agent in the disease, nor was it until Mr. J. Hume Patterson, Assistant Bacteriologist to the Corporation of Glasgow, discovered the true agent to be a micro-organism now known as *Bacillus salmonis-pestis*, that *Saprolegnia* was recognised as fulfilling no more than its ordinary function of spreading as a mould upon dead animal matter; those portions of the salmon's muscular tissue as had been necrosed (or killed) before the fish left the sea by the bacillus affording it a suitable nidus. So it is with *Botrytis*; if one can keep any lily in vigorous growth and health, there is not the slightest cause to apprehend attack from the fungus. There are, of course, many enemies, chiefly subterranean, such as mice, voles, slugs and grubs, which cause the disappearance of bulbs that might otherwise prosper satisfactorily. These are not taken into account here, for I am treating only of those lilies which have refused to respond to the most considerate treatment in our west country garden. Of these, the chief is *Lilium Browni*—chief, because on the whole I think it is the loveliest of all lilies, relying for effect not upon brilliant colour, but on a quiet scheme of ivory white and purplish maroon, punctuated by heavy anthers of rich russet (Plate III). It is the perfect form of the flower that makes it peerless

Flowers

in its kind. This splendid lily has been rightly classed as difficult. " Sometimes," says Mr. Grove in his useful handbook, " it does well for a time in light, sandy soil, raising the hopes of the grower only to dash them down again by dwindling away in such soil."[1] That is exactly our experience here. Some bulbs, planted on a steep slope facing south and ensuring thorough drainage, were tilted on one side to prevent water lodging in the cupped scales. They flowered generously the first year, grudgingly the second, and nothing appeared above ground in the third. Giving them up as lost, we planted a nice bit of *Osmanthus Delavayi* on the spot whence they had vanished ; and lo ! in the fifth year up shot a strong flowering stem, bearing a single flower of such exquisite beauty as to lure one on to fresh endeavour. Ah me ! when I remember the luxuriant batch of this fair but frail one that used to breathe its fragrance through the late Sir Henry Yorke's woodland at Ivor Heath. Does it do so still ? I wonder.

Concerning *Lilium Humboldti* Mr. Grove utters the discouraging warning that it is eminently a lily for the specialist.[2] Timely warning this, for one who claims no higher grade than that of a diligent amateur ; nevertheless, I have devoted as much attention to this wayward beauty as might, if applied to a less elusive aim, have secured for my heirs a comfortable competence. Ravages of *Botrytis* after a single season of display indicate unmistakably a constitution undermined by unsuitable environment. The gay little

[1] *Lilies*, Gardening of To-day Series, p. 80. [2] *Op. cit.* p. 86.

Some Failures

L. Philadelphicum and the stately *L. Parryi* also have refused to respond to the most ardent solicitude for their welfare; likewise the rosy-flowered *L. rubellum*, but with the last-named, I suppose, my experience has not been much more disheartening than that of most people who have tried to establish it. No doubt the best chance of doing so is one that we have neglected, namely—to raise it from seed.

This is, I think, the shortest chapter in the book; whence it must not be argued that our list of failures is here complete. Only those are recorded which occur to mind when writing; but there must be many other fair flowers that have disappeared from our borders without any obituary notice:

> They pass forgotten, as a dream
> Dies at the opening day.

Whereas failures such as I have enumerated serve to enhance the gratification assured by success with some new and rare species, let me relieve the gloom of this catalogue of disaster by recounting our experience with what has proved to be, in my esteem at least, the fairest of all pæonies. Its name is *Pæonia Cambessodessi* (Plate XII), and it came here a few years ago as a gift from a generous Irish lady, Miss Geoghegan, who found it growing on the sea-cliffs of the Balearic Isles. It is the earliest of the family to flower, bearing its beautiful rosy cups throughout April. It is only about 12 or 15 inches high, and the leaves remain fresh throughout summer and autumn, shining bronze on the upper surface and bright red beneath—a truly delectable little herb.

X

Some Weeds

O grief for the promise of May, of May?
O grief for the promise of May!
Tennyson.

DR. JOHNSON in his dictionary defined a weed as "an herb noxious or useless." One hundred years later Professor Skeat explains it as "any useless and troublesome plant." Neither definition is quite satisfactory from a gardening point of view, forasmuch as there are many plants which are admirable in positions suitable for them, but which prove a terrible nuisance in a place intended for something else. Just as dirt has been explained as matter in the wrong place, so a comprehensive definition of a weed is a plant where it is not wanted. Some of the most troublesome weeds in other countries are carefully cultivated for their beauty in our gardens, the *Lantana*, for instance, which is said to be a holy terror in the West Indies. It is not my purpose to treat of what every gardener recognises as weeds—dandelions, plantains, groundsel and the like—but to warn others against incurring the trouble which I have brought

PLATE XII

⅜ natural size   PÆONIA CAMBESSODESSI   4 May, 1918

Some Weeds

upon myself by planting in the flower borders certain exotics of uncontrollable vagrancy and greed.

Except the *Ranunculaceæ*, there is no more Protean family of plants than the knotweeds—*Polygonaceæ*, and none that demands greater discretion in planting, for it contains terrors as well as treasures. Chief among the terrors are *P. Sachalinense* and *polystachyum*, and timely warning may be of service to those who have not yet encountered them. No parents ever exercised more anxious care in starting a boy in his career than we did when scraps of these plants arrived here as novelties; we fixed upon the choicest nooks to harbour them—nooks which they treated merely as a starting-point for distant excursion. Their subterranean runners spread to an amazing distance; they pass under gravel paths as featly as an eel glides through mud, and throw up their ample foliage on the other side on their way to still further conquest. For conquest it is. They throw a shade under which no green thing may survive; they suck up nutriment from the soil that should go to maintain their betters.

It cannot be denied that both these species are of remarkable beauty. *P. Sachalinense* will only develop its fine foliage—broadly ovate leaves a foot long—in full exposure, and looks splendid flapping them gently by the waterside, especially after they have turned bright yellow in October. But even in such a position it should be kept in a distinct clump, else it will soon grow into a featureless thicket. The flowers of *P. Sachalinense* are not of much account; but the

Flowers

feathery panicles of *P. polystachyum* are like lovely lacework, and in a woodland where this species may be suffered to take its fling have a charming effect until cut short by frost. For several reasons I am glad that we have not planted *P. cuspidatum*, as rampant a robber as either of the others, one being that, having already drawn so heavily on the vocabulary of vituperation upon its kindred that there is no expletive left of sufficient virulence to greet it withal. *P. campanulatum* is a newcomer from the far east, with really beautiful flowers—rose and white. It is only in its second year here, and although it is spreading, it does so above-board, so to speak, and looks as if it can be kept within reasonable bounds. It is a strange thing that rabbits will not touch the succulent shoots of any of these knotweeds.

Of the species of this family that may be safely planted in the garden proper, *P. Baldschuanicum* and *Auberti* are rapid climbers, well worthy of placing where they can run over a trellis, a ragged pole or a tree stump, which they will veil with sprays of white and pinkish blossom. *P. Alpinum* can go in the herbaceous border, giving pretty white flowers in June, while the quaint *P. equisetiforme*, after posing as a clump of rushes all summer, decks its two-feet wands in early September with numbers of sessile starry white flowers. It is not very hardy, but is good enough to deserve good shelter. Our native *P. bistorta* may be given two or three square yards to good effect if the ground is fairly moist; *P. sphærostachyum*, from the Himalaya, is an alpine gem which we have hitherto

Some Weeds

failed to keep; but *P. vaccinifolium* from the same region is a trailer which covers part of a west-fronted retaining wall in autumn with sheets of bright pink heath-like blossom, a really good thing.

The extraordinary variety and dissimilarity among species of the knotweed family has led me astray from the purpose of this chapter, which is to pillory those plants which it is expedient to exclude, or if that be not possible, to expel from the borders and rockwork. Let me get back to the subject by naming that lively little Mexican, *Erigeron mucronatus*, sometimes still known as *Vittadenia triloba*. Its behaviour here has tested, if it has not strained, my friendship for the kind lady who gave me a slip of it some years ago. Inserted in a crevice of a retaining wall, it delighted us during two or three seasons with its crowds of daisy flowers, which open pink, turn white, and die off red, a pleasing trick prolonged through summer and autumn, were it not that these flowers were followed by seeds, which get transported somehow to places where they are not wanted. Fresh plants spring up; the process is repeated over and over again, and the utmost vigilance fails to detect the seedlings before they have sent wiry roots of extraordinary strength so far among the stones that it is impossible to extract them without pulling the wall to pieces. Soon they form a dense mat, smothering any helpless alpine that lies in their path. The proper use for this Mexican daisy is to drape old buildings or walls well outside the garden. It has a pretty effect on the wall skirting the high road between Cintra and Collares in Portugal.

Flowers

Nature has shown singular perversity (from a gardener's point of view) in assigning perennial existence to our native Welsh poppy—*Meconopsis Cambrica*—whereas all the beautiful Indian species—nearly all, at least—are only biennial.[1] The Welsh poppy, indeed, is a pretty flower, especially when the orange variety grows with the ordinary yellow type in woodland; but, unluckily, it seldom gets a chance in our woods, rabbits being especially partial to it. In a flower garden it can best be described in the words of the prophet Daniel as " the abomination that maketh desolate "; wherefore let no one who values his peace introduce it, as I did in a fatuous moment, many years ago, since which life has been one long and costly warfare. Costly—because this plant has the knack of depositing its seeds in the very heart of some precious perennial, where, if the seedling is not detected in infancy, it anchors itself with a tap root so long and strong that it is impossible to eradicate it without lifting the other. It is especially fond of making for its offspring a nursery among the rhizomes of bearded iris, which resent disturbance and are apt to repay it by not flowering in the following year. The Iceland poppy—*Papaver nudicaules*—behaves much more reasonably, yielding much the same effect of colour, and deserves all encouragement to sow itself through the borders.

Campanula lactiflora is one of the handsomest bell-flowers and a most desirable ornament for woodland, where it throws up stems five or six feet high,

[1] *Meconopsis simplicifolia* sometimes flowers in a second and even a third season.

Some Weeds

richly plumed with panicles of bloom varying in colour from white to violet-blue. But let no one who sets store by ease of body or peace of mind allow it a place in the flower garden. I met this fine plant first at Belvoir, and esteemed as good booty some pieces that I brought thence; but I have lived to repent treating it as a garden flower, so lavishly does it scatter its seeds, so vigorously does it dominate other plants, and so prodigiously deep does it send its roots.

It was in an evil hour when we introduced a certain *Oxalis* into the garden. I do not know the species; it has handsome leaves barred with purple, abundant out of all proportion to its pinkish-lilac flowers. It spreads with terrific rapidity, and it is almost impossible to get rid of it. *Anemone Japonica*, beautiful as it is, requires discipline lest it over-run things of equal merit; and as for *Veronica ambigua* it has been the cause of a good deal of language more forcible than polite.

Cynoglossum amabile is a recent introduction from China, and is certainly one of the loveliest of hound's-tongues. It is only biennial, but renews itself liberally from self-sown seed. It is the nature of those seeds that compels me to include this fine plant in the category of weeds, for the seeds are contained in burrs of a peculiarly prehensile kind, which cause intolerable irritation to any dog or cat that comes in contact with them. It is impossible to rid the animal of them except by cutting the hair or fur. Under a lens the complex structure of these burrs appears very interesting. For the same reason we have done our utmost to

Flowers

expel two or three species of the New Zealand *Acæna* from borders and rockwork. It may be objected that neither cat nor dog has any business in such places. Granted: but the suffering inflicted by the burrs of these plants is too severe punishment for occasional trespass, especially as the cat is supposed to go to such places in pursuit of voles and mice.

The South African genus *Tritonia*, usually known among gardeners as *Montbretia*, is generally described as only half-hardy, but in west country gardens *T. Pottsi* and its hybrids have established themselves so aggressively that we have difficulty in keeping them within bounds, and they are well within the category of border weeds. The flowers are showy, in many shades of yellow, orange and red, and as rabbits do not fancy the foliage, *Tritonia* may be trusted to take care of itself in woodland and waste places, where it spreads fast and lights up its flames in latish summer. The kindred *Antholiza paniculata* is not so rampant, and also makes a fine ornament for the wild garden.

Asters of the Michaelmas Daisy class need to be closely looked after. We cannot afford to dispense with their cheerful illumination of the waning year; but they spread so fast, both by runners and seed, that they soon render a border uninhabitable by plants of less vigour unless they are kept rigorously in check. And a border filled with nothing but asters is a dull affair. To include *Anemone Japonica* among garden weeds may seem an unmerited affront upon a beautiful flower; but it is also an aggresive one, at least in those soils which are to its liking. In some gardens

Some Weeds

it is difficult to get it to thrive at all; in others, it rambles far, and is apt to suppress things of less vigour. Of like nature is the willow gentian—*G. asclepiadea*, a plant whereof one cannot have too much in the woods; but it produces such a bumper crop of seed that strict discipline must be enforced to restrain it from claiming too much space in the borders. Far more difficult to keep within bounds are some species of *Helianthus*—I think *H. multiflorus* is the correct title of the kind which has given us most difficulty. Some of the Meadowsweets require careful herding, so widely do they spread their seeds. Among them the handsome goat's-beard—*Spiræa aruncus*—and the pretty Queen-of-the-Prairie—*S. lobata*—are charming flowers, but very capable of overriding weaker things; and as for the ruddy purple *Astilbe Davidi* from China, it is a perfect pest, taking exclusive possession of wide tracts of any garden soil moist enough for its fancy. Woodland—and waterside woodland for choice—is the proper home for it, where it may display its five-foot crimson plumes with their singular azure sheen to much advantage.

Those who have been seduced by the pretty pink blossoms and silvery foliage of *Convolvulus althæoides* to admit it to the flower borders, must bitterly have rued the deed, for it is a restless and irrepressible rambler. Yet it is so fair in form and colouring that one cannot be cross with it; and I cannot but regard it with affection from association with the gentle châtelaine of Patshull, from whose garden I received it in a present.

XI

Some Plant Names

> Who hath not viewed with rapture-smitten frame
> The power of grace—the magic of a name?
> *Thomas Campbell.*

IT is many a year since old Wyber, gardener to the late Sir David Erskine of Cardross, was gathered to his fathers. He was master of his craft and was gifted with a quiet sense of humour, and I am reminded of one of his quaint sayings as often as I hear people grumbling, as they often do, about the Latin and Greek polysyllables employed to denote garden flowers. Such names caused no perplexity to Wyber. A visitor to his garden once asked him how in the world he managed to remember them.

"Oh," he replied, "I mind them easy enough. Ye see, I hae what's called a memoria technica for them."

"Memoria technica," said the visitor; "how does that work?"

"Weel enough, sir, weel enough," said Wyber. "There's a tree now, the *Cryptomeria Japonica*. When I'm teaching the lads I tell them when they look at

Some Plant Names

that tree to say—'Creep to the mear[1] and jump on to her.' Ane o' the lads was gaily surprised when I was giving him a lesson in the greenhouse ae day. I was showing him the *Selaginella stolonifera*—a kind o' moss, ye ken. If ye want to mind the name, says I, think o' Silly Jean and Nellie stole a neive-fu'.[2] 'Losh! Mr. Wyber,' quo' he, 'them's my twa sisters.' "

My old friend Mr. William Robinson, who must have forgotten more about hardy plants than most of us have ever learnt, and who has done more by precept and example than any other writer of our time to restore them to favour, is a strenuous advocate for the use of English names for them, and has expressed impatience with the practice of coining titles for them in dead languages. He would have us speak of saxifrage as rockfoil, of *Clematis* as virgin's bower, of *Anemone* as windflower, of *Aubretia* as rockcress.

"The stock of bad Latin," he complains in the preface to his useful little book on the clematis, "which we owe to the botanists leads some people to cut capers with that language with fearful results; the terms which, issuing from the mouths of botanists are bad enough, when descending into those of gardeners are grotesque indeed. In botany these technical terms may be essential, but gardening is quite a different affair, and for ages the effect of botanical classifications on the garden has not been a happy one. Nor are they necessary; the names in our own tongue are as good as any, and we are not prevented from adding the Latin name when necessary."

That is all very well, so far as it goes; but it does not go far enough even for practical garden use.

[1] Scottice for "mare."   [2] A handful.

Flowers

English names could not be minted fast enough to fit a tenth part of the new species being introduced continually from foreign lands. Even if they could be so minted they would not pass current among other nations than ours, wherefore botanists and gardeners long ago found it expedient to classify plants under a nomenclature framed in a dead language, in Latin at first as the common tongue of science, and later in Greek. Let me not be suspected of any slight upon such English plant-names as we have, most of them hallowed by antiquity, and many of them spontaneously poetic. He would be a sorry pedant who should speak of lily-of-the-valley as *Convallaria* or of rose campion as *Lychnis diurna*. Well said Thomas Carlyle in one of his less mordant moods—" Giving a name, indeed, is a poetic art; all poetry, if we go to that with it, is but a giving of names." But we have to be practical as well as poetic; it may be more patriotic to speak of rockfoil instead of saxifrage, notwithstanding that the latter is acknowledged as good English in all dictionaries, Dr. Johnson's included though he never heard of rockfoil.[1] After all, the main purpose of a name is identification; the English language has become the vernacular in lands far outside the bounds of Great Britain; our fellow-countrymen of the overseas dominions and our kinsmen of

---

[1] Under the word "saxifrage" Dr. Johnson quotes Quincy to the effect that the plant was so named—*quasi saxum frangere*—in the belief that it was a remedy for stone in the bladder. The doctrine is cited by Pliny: *calculos e corpore mire pallit* (*Nat. Hist.*, part i. lib. xxii. cap. 9). It was naively supposed that, because the plant grew in clefts of rock that it would cleave stone in the human subject.

Some Plant Names

the United States use many English names to denote plants in which they see a resemblance to others in the old country, although these may belong to quite different natural orders and genera. Nay, even in our right little, tight little island the popular name of a plant in one locality is used to denote quite a different one in another. When a Scots forester speaks of a plane he does not mean the tree which contributes so much to mitigate the murkiness of London parks and square—*Platanus orientalis*—but the sycamore—*Acer pseudo-platanus*. Dr. Jamieson in his Scots dictionary interprets "craw-taes"[1] as "crowfoot," *Ranunculus repens* and *acris*, the equivalent of the English "crowfoot"; but in some parts of Scotland it means the bird's-foot trefoil—*Lotus corniculatus*—and in others the blue hyacinth.

It may be objected that in these instances the confusion arises out of our inability to speak decent English in Scotland. I accept the taunt and go for an illustration to the "well of English undefyled."[2] Every country-bred child, from the Land's End to John-o'-Groat's, knows what flower is meant by forget-me-not—Hood's "blue, significant forget-me-not"—but our great-grandsires understood it in a very different sense. All the English herbalists and botanists, from Lyte and Gerard in the sixteenth century down to Samuel Gray, who published his *Natural Arrangement of British Plants* in 1821, applied

[1] Crow's-toes.

[2] Not to Dan Chaucer, whom Spenser apostrophised, but to the language itself.

Flowers

this name to the yellow bugle or ground-pine—*Ajuga chamæopitys*[1]—which had earned it from the disagreeable taste which, like some others of the Labiate order, it leaves in the mouth when bitten. *Myosotis*, which we now call forget-me-not, was always known as scorpion-grass, from the fanciful resemblance of the raceme of unopened flowers to a scorpion's tail. Henry Lyte (1529-1607) states that "it hath none other knowen name than this," and so this hairy annual continued to be called in England until about the middle of the nineteenth century a poem was published (forgive me for forgetting by whom) describing how a lover was drowned in the endeavour to pull some of the marsh *Myosotis* for his mistress, flinging it to her as he was swept away by the river with the prayer "Forget-me-not!" Again, one writing to an English nurseryman for a hundred bluebells would assuredly receive that number of wild blue hyacinth bulbs; whereas a similar order to a Scottish firm might be interpreted as signifying what are known in the south as hare-bells.

Some people are deterred from taking an interest in all but common garden flowers by what they call the long-tailed unmeaning names attached to less familiar plants. George Crabbe, combining in his own person the parts of botanist, parish priest and poet, himself owned an arboreal surname and waxed mildly satirical on scientific nomenclature.

---

[1] The common name for this weed, "ground-pine," is itself an illustration of the uncertain character of popular terms, for the only quality that connects it with the pines is its resinous odour.

Some Plant Names

"Why *Lonicera* wilt thou name thy child?"
I asked the gardener's wife in accents mild.
"We have a right," replied the sturdy dame;—
And *Lonicera* was the infant's name.
If next a son shall yield our gardener joy,
Then *Hyacinthus* shall be that fair boy;
And if a girl, they will at length agree
That *Belladonna* that fair maid shall be.
High-sounding words our worthy gardener gets,
And at his club to wondering swains repeats;
He there of *Rhus* and *Rhododendron* speaks,
And *Allium* calls his onions and his leeks;
Where Cuckoo-pints and dandelions sprung
(Gross names had they our plainer sires among),
There *Arums*—there *Leontodons* we view,
And Artemisia grows where wormwood grew.
But though no weed exists his garden round
From *Rumex* strong our gardener frees his ground;
Takes soft *Senecio* from the yielding land,
And grasps the armed Urtica in hand.[1]

The justification of using the dead languages for classifying plants and animals consists, first, in the fact that they *are* dead, and therefore subject to no change in form or meaning, and second, that they are the common property of all civilised nations. Greek and Latin, therefore, afford the surest medium to secure precision. Even so, it requires some vigilance not to be led astray. My old friend and correspondent, the late Canon Ellacombe, will be long remembered, not only as a keen horticulturist and learned botanist, but as a delightful host in his vicarage of Bitton. He was also a classical scholar of no ordinary erudition; but no man is infallible, and he strangely missed the

[1] *The Parish Register*, Part i.

Flowers

significance of Falstaff's allusion to "eringoes" on meeting Mrs. Ford and Mrs. Page in Windsor Park.

"Let the sky rain potatoes: let it thunder to the tune of Green Sleeves, hail kissing comfits and snow eringoes."[1]

In his *Plant-Lore and Garden-Craft of Shakespeare*, Canon Ellacombe, after noting that Gerard explained "eringoes" as the candied roots of the sea-holly—*Eryngium maritimum*—proceeds:

"I am inclined to think that the vegetable Falstaff wished for was the globe artichoke, which is a near relative of the eryngium, was a favourite diet in Shakespeare's time, and was reputed to have certain special virtues which are not attributed to the sea-holly, but which would more accord with Falstaff's character."

Now, herein the learned Canon begins by making a slip in botany, for the globe artichoke and the sea holly are far from being near of kin, the first belonging to the *Compositæ* or Daisy order, the second to the *Umbelliferæ* or Hemlock order. He goes on to miss entirely the point of Falstaff's mention of "eringoes" in his list of incentives to amativeness. Whether Shakespeare referred to what we know as the potato—*Solanum tuberosum*, Linn.—or to the sweet potato commonly grown in the West Indies—*Convolvulus batatas*, both were believed by physicians of the sixteenth century to possess aphrodisiac properties; and for several centuries the eryngo, whereof our native sea-holly is a familiar species, has borne a similar reputation. But it is certain that it earned that reputation through being confounded with a cruciferous plant bearing the somewhat similar name of *Eruca*.

[1] *Merry Wives of Windsor*, Act v. Sc. 5.

Some Plant Names

The N.E.D. gives Falstaff's speech, above-quoted, as the earliest occurrence of "eringo" in English literature, the date of the *Merry Wives of Windsor* being 1598. Pliny, in discoursing of *Erynge sive Eryngion*, mentions it as an effective remedy against the venom of snakes and other poisons,[1] but has not a word to say about it as an aphrodisiac, nor do I know of any Latin or Greek writer who attributes stimulant properties to this herb. On the other hand, Pliny writes confidently of *Eruca* as *concitatrix Veneris*[2] and many other writers refer to that plant in similar terms. It is one of the Cruciferæ, known in botany as *Eruca sativa*, a native of the Mediterranean region. Ovid in his *Remedium amoris* warns those who would avoid falling in love against eating it :

Nec minus erucas aptum est vitare salaces—

and Martial recommends it to be given to husbands negligent of their wives :

Concitat ad venerem tardos eruca maritos.

Columella prescribes sowing it near the effigy of Priapus in gardens. It is clear, I think, that before the close of the sixteenth century *Eruca* had been confused with *Eryngo*, the stimulating properties of the former thus coming to be ascribed to the latter. Nor is there any reason to suppose that the mistake mattered in the least, the prescription of either herb as a drug being sheer quackery.

Instances have occurred of a plant sailing under false colours, so to speak, through a mistaken report

[1] *Nat. Hist.*, lib. xxii. cap. 7.   [2] *Ibid.* lib. xix. cap. 8.

Flowers

of its origin. Such is the case with what may be seen in some nurserymen's lists as *Saxifraga cotyledon Icelandica*. That fine variety was received by an experienced amateur in Oxford from a friend who, he understood, had collected it when travelling in Iceland. The Oxford gentleman very kindly gave me a piece of it, and it was not until after it had pleased us all with its great sprays of snowy flower that I was informed that it was in Norway that the collector had spent his holiday and that he had been misunderstood to say that he had been to Iceland.

Plant labels of the soft wood usually supplied by seedsmen are most treacherous things, fit for nothing but marking sowings of annuals. They are worse than ever now since it has become the practice to mix barium with white lead, with the result that the paint vanishes, and with it the inscription, with the first frost followed by wet. The late Sir Isaac Balfour gave me from the Edinburgh Botanic Garden a species of aster under the name of *A. Lipskii*. It throve and flowered, revealing itself as a starwort of much merit, bearing large solitary blossoms in July on stiff eighteen-inch stems. Desiring to know something about the origin of this plant, I wrote asking about it. Sir Isaac being no longer in the post which he had occupied for so many years with such distinction, his successor in office told me that nothing was known of the *provenance* of the aster, and that the title it bore was probably a ghost name, arising out of the illegibility of a wooden label!

One of the last—nay, I think it was the last—of

Some Plant Names

many wrinkles imparted to me by my old friend H. J. Elwes, a few weeks before his death, was a recommendation of holly as the best material for wooden labels. We happened to have a large log of holly in the wood yard; it was sliced up into suitable slips, since which, to quote a classic in advertisement, " I have used no other." The beautiful, smooth, hard and white surface of this wood enables one to dispense with paint altogether; it may be inscribed with an indelible pencil.

Precision being the cardinal requisite in the scientific names of plants, they ought, so far as possible, to be either commemorative, as is *Berberis Darwini*, the barberry of Charles Darwin who discovered the species, or descriptive of the plant. As a rule, specific names conferred by botanists answer one or other of these requirements well enough. A morning spent in bitter conflict with weeds in the flower garden, ending in my withdrawing *not* according to plan, brought to mind how aptly Linnæus had named that furtive pirate *Ranunculus repens*, which had taken up its quarters in a clump of the charming little Spanish hyacinth (not the more robust *Scilla Hispanica*, but the slender thing whereon Linnæus bestowed the name of *Hyacinthus amethystinus*), whence it had shot out its vigorous runners, which, rooting at frequent intervals to found a fresh colony, would in time have carpeted the whole border. Now in naming this hyacinth Linnæus must have had before him a discoloured specimen from a herbarium, for there is no trace of amethyst purple in a fresh flower of this graceful plant; it is sky-blue of the same tender tint

Flowers

that we have in *Mertensia Siberica*. The Greeks named the purple gem ἀμέθυστος—" not drunken "— because it was reputed to be a cure for drunkenness; but Linnæus can hardly have imputed that virtue to this little hyacinth. It is not easy to find him astray in the epithets he used to denote species. The more closely one contemplates the scope and detail of his work in classifying the animal and vegetable kingdoms, bearing in mind the difference between his day and ours in the matter of transit and communication, the more reverently must one recognise the grasp and penetration of a commanding intellect. Modesty is not an invariable attribute of genius; but it was no affectation of humility that prompted Linnæus when he conformed with the custom of botanists to attach their names to certain plants. Having at his disposal the whole host of green things that he was the first to array in order, he chose no lofty pine, no massive oak, no gorgeous flower; it was the fragile, creeping *Linnæa borealis* that he appointed to bear his name and remind future generations of the work he had accomplished for them.

Although the classification of plants devised by Linnæus has been superseded by the adoption of the Natural System, his scheme of nomenclature, consisting of a single generic and a single specific name, has stood the double test of time and advancing science. Despite the enormous additions to the lists of genera and species, it still holds the field in virtue of its precision and simplicity. Some confusion became inevitable as time went on, owing to the

Some Plant Names

absence of any central authority to regulate the naming of new species. Synonyms became increasingly frequent, wherefore at the International Congress of Botanists, held at Brussels in 1910, a committee was appointed to consider the best means of securing uniformity of nomenclature. Its chief recommendation was that thenceforward every genus and species should bear the name applied to it by the botanist who first described it. Of minor importance was a resolution to the effect that when the specific name of a plant is formed from that of a person or place, it should be written with a capital initial.

To these rules it behoves every good gardener and amateur to conform; but custom dies hard, and it is with some reluctance that one discards an euphonious name for one that may be much the reverse. For instance: having long known a fragrant evergreen as *Oreodaphne Californica*, meaning the Californian mountain laurel it costs one an effort to think of it under the ungainly name *Umbellularia*, which means we know not what. The African torch' lilies or red-hot pokers have undergone singular vicissitude of nomenclature at the hands of successive botanists, the genus having been successively named *Rudolphæmeria, Triclissa, Tritomanthe, Tritomium* and *Tritoma*. Every gardener had learnt to know it under the last of these names, until, under the rule of priority, it became *Kniphofia*, because, two hundred years ago, a German professor of medicine had affixed to it his own name—Kniphof.

Then there are a few specific names with historical

Flowers

association, such as that which Sir Joseph Hooker gave to the noblest of all Indian rhododendrons when he discovered it and christened it *R. Aucklandi* to commemorate the viceroyalty of Lord Auckland, albeit that was not one of the happier phases of British rule in India. Hooker was not aware that this rhododendron had already been discovered and named by an explorer named Griffith, whose priority has now been recognised by fixing upon it the awkward name of *R. Griffithianum*. Howbeit, honour to whom honour is due; one must not grudge it to pioneers in science, though we may wish that they had always contented themselves with the simple possessive case instead of coining cumbrous adjectival forms. *R. Griffithi* would be a more manageable mouthful than R. *Griffithianum*. Such names, if they are to serve the purpose of commemorating individuals, ought not to be pronounced as one often hears them, in a way that voids that purpose. *Rhododendron Fàlconeri* and *Hòdgsoni* should be spoken with the stress as it is here marked by an accent, not *R. Falconeèri* and *Hodgsôni*. *Fuchsia* is universally sounded "fewsha" thereby disguising it as reminiscent of Leonard Füchs, a diligent botanist who died in 1566.

Still, it is all fairly plain sailing so long as the surnames chosen to denote species, and Latinised for that purpose, are those of reasonable simplicity. *Lilium Browni* and *Rhododendron Thomsoni* are names conveying little suggestion of the fragrance of the former or the brilliance of the latter, but at least they are easily pronounced and not difficult to remember.

Some Plant Names

But when Polish and Russian botanists are to be commemorated by giving them floral namesakes, it requires the gift of tongues to master the harsh polysyllables that are evolved. In my youth there was a game played with ivory letters, which were served out in batches to the players, who were bidden to find out the word which each batch contained. I was reminded of the sort of hopeless concatenation the letters used to fall into when I came up against such plant names as *Pæonia Mlokosewitchii* and *Gentiana Przewalskyi*!

"Obstupui, steteruntque comæ et vox faucibus hæsit."[1]

Perhaps it does not become one whose native tongue is English to criticise the speech of other nations, seeing that of all people who on earth do dwell, those of Teutonic race show least regard for agreeable sound in personal names. We have distorted the Norman Lizavir into Elizabeth; Diana as we pronounce it is a harsh rendering of the liquid Italian, and we have allowed the shrill Jane, which was never heard among our people till the 16th century, to silence the softer Joan. Nor do we show a sense of music when we invent pet names. The only alternatives we provide for Thomas is Tom or Tommy, rude equivalents to the Italian Masaccio or Masaniello. But the Italians speak the most beautiful language on earth. One wonders what form they would have given to the name inherited by the descendants of John of Sevenoaks, whom we know as the family of Snooks. Some years ago, an English couple, good

[1] I was bewildered, my hair stood on end and my voice stuck in my throat.

Flowers

friends of my own, were living in Italy; the valet in their establishment answered to the name of Duval, and the lady's-maid to that of Philomena Orsini. Letters looked very attractive when they arrived addressed—

> Alla gentilissima signora
> Philomena Orsini.

Travelling in the East, I had a dragoman named Giovanni Battista, which somehow sounded more imposing than the English equivalent—John the Baptist.

Still, be the shortcomings and asperities of our English vernacular what they may, our forefathers managed to coin out of it some endearing names for common flowers, which, hallowed as they are by association and mellow with age, it would ill become us to allow to fall into disuse. It is refreshing as well as instructive to spend half-an-hour with Dr. Prior, turning over the pages of his *Popular Names of British Plants*.[1]

In one respect we owe a debt of gratitude to the International Committee in that they have steered clear of the maze of absurdity into which a certain school of ornithologists have blundered. To cite a single instance—Linnæus named the common blackbird *Turdus merula*; when it was decided, doubtless for good anatomical reasons, to class thrushes and blackbirds in separate genera. It was decreed that the blackbird was in future to be known as *Merula merula*. Nor did these reformers stop there, for one

---

[1] London, Frederic Norgate, third edition, 1879.

Some Plant Names

may read now in publications conforming to the new scheme about *Merula merula merula*! Whether this is meant to denote a variety from the typical blackbird, I know not; but it sounds uncommonly like classification gone dotty. Bitterly did old Tegetmeier deride what he called the *Asinus asinus* system. That there is need for some discretion on the part of botanists also appears from the recognition in the Kew *Hand List* of a variety of the little bulbous *Chionodoxa Luciliæ* as *gigantea*. Shade of Linnæus! a plant not more than six inches high to be distinguished as gigantic. And one must hold in leash all sense of humour in obeying the decree that a little creeping bell-flower, long known as *Campanula muralis* (a most apt name for a plant that clothes dry walls) shall henceforth bear the sounding title of *Campanula Portenschlagiana*!

In a few cases the scientific name of a plant has fitted itself conveniently into popular English, and one speaks colloquially of geranium, arum, asparagus, clematis, kerria, ribes, etc. Rhododendron, however, is a cumbrous polysyllable which we should do well to discard for the simpler American name "rosebay."

We speak complacently of a cherry and a pea, names which have become far too deeply rooted in modern English to be dislodged; but both of them have their origin in a misconception. "Cherry" we inherited or borrowed from the French *cerise*, which looked like a plural noun, so we cut off the final sibilant to represent a single cherry. Precisely the same process has evolved the name "pea." The Romans

Flowers

having borrowed their *pisum* from the Greek πίσος, it appears in Anglo-Saxon as *pisa*, plural *pisan*, whence the Middle English *pese*, plural *pesen* and *peses*, and it remained for later generations to create the ghost-name " pea " for the singular number. We speak of pea-soup, which is well enough, but how are we to express the plural of a sweet-pea ? Sweet peasen and sweet peases might sound pedantic; there remain the alternatives of sweet-peas or sweet-pease. In that connection it may be noted that, although it has been reported that the Scottish clergyman who, a few years ago, won the *Daily Mail* £1,000 prize for the best sweet peas, did on the following Sunday, in giving out the hymn " Peace, perfect peace " pronounce the words with a soft sibilant, we are assured that the story rests upon a very uncertain foundation.

XII

L'Envoi

I CANNOT bring these rambling pages to a close without putting on record how greatly the interest of our garden has been enhanced by gifts from friends. To these, also, borders and shrublands owe a large measure of such beauty as they possess; but the chief value of such gifts consists in the memories with which they are fraught. A stroll round the garden on a summer evening would be reft of more than half its charm were it not that at almost every turn some well-remembered features are brought to mind—some familiar accents offer greeting. Those milk-white sprays of *Libertia* take me back to the distant day when Miss Frances Hope of Wardie Lodge, Edinburgh, gave me a piece of it whence all our plants are descended. She died three-and-forty years ago, but I have but to close my eyes to behold her kindly, sun-tanned face under the lilac sun-bonnet that crowned her working dress of short skirts, soiled gauntlets and heavy shoes as she led me round her borders. And there is gentle old Dr. Lowe in the

Flowers

black frock coat which he was too eager to change after release from professional duty before hurrying into the delightful garden he had furnished on the west side of Edinburgh.[1] He is bending over a plant of mandrake—*Mandragora officinarum*—that harmless but uncomely herb so fraught with myth and superstition, which has flourished here ever since I received it from him in the 'seventies. Then there is an amiable weed of the order *Carophyllaceæ* whereof I know not the name, and, as it is not in flower at the moment of writing, I cannot tell whether it is a *Dianthus*, a *Lychnis* or a *Silene*. Anyhow, it has been allowed to run at will through our borders ever since I stole a pinch of its seed from Lady John Scott's borders at Caroline Park (now, alas, an ink-factory) near Granton. Her ladyship was an energetic antiquary, and once fired the heather—metaphorically—by digging up a plague pit in a populous district!

No country laird ever had a better or truer neighbour than I had in the person of the tenth Earl of Stair. He died in 1903, but there stand two Lawson cypresses of Fraser's fastigiate variety, now five-and-twenty feet high, one on each side of a flight of steps leading off the terrace, which he gave me as plants a foot high the year before we lost him. The *Allium* family is not commonly associated with sentiment, but that group of *A. sphærocephalum* is endeared to me by reason of the original root having been pur-

---

[1] I cannot recall the name of Dr. Lowe's house near Edinburgh. After resigning his appointment, he migrated to Wimbledon, and lived long enough to create another most interesting garden there.

L'Envoi

loined from the flower garden of Mary, Countess of Galloway, whose bright spirit was released from this perplexing planet just twenty years ago. That little rose bush with single crimson flowers is *Rosa Indica*, the wild plant whence have descended all the Chinese monthly roses. It was part of the spoil with which Canon Ellacombe loaded me during my first visit to Bitton Vicarage. Not far from it is a plant of fairy delicacy—*Alstrœmeria Hookeri* (Plate XII)—which recalls the commanding figure and sounding voice of my old school-fellow, the mighty traveller, H. J. Elwes, who discovered it in the pass through which Darwin crossed the Andes. Much of the value which I attach to a good bush of *Rhododendron Loderi* in the wood skirting our garden derives from the fact that it was a gift from Sir Edmund Loder who, by crossing *R. Fortunei* with *R. Griffithianum*, raised at Leonardslee this, the most gorgeous hybrid in the genus. Shrubs and herbs without number serve to revive memories of that great botanist and generous friend, Sir Isaac Bayley Balfour, who once told me that he considered his main duty as Director of Edinburgh Botanic Garden was to distribute plants among those who were likely to grow them successfully.

All these old friends have passed behind the veil :

Lange leben heisst viele überleben—[1]

but our borders and woods are enriched with countless fair things to remind us of others who are still busily

[1] To live long is to outlive many.—*Goethe*.

Flowers

alive. No costly jewelry, no gift-book, no picture serves so surely as a living bush or herb to bring an absent friend to mind, and life without friends would be more desolate than a garden with neither flowers nor fruit.

THE END.

General Index

American Wood-Lily, 138.
Andalusia, 169.
Anemone, 2.
Aralia, 100.
Ardgowan, 3.
Arncliffe Valley, 39.
Arnold Arboretum, 65, 116.
Australian Bottle-Brushes, 88.
Autumn Crocus, 10.
Avebury, Lord, 41, 58, 98.
Azalea, 102, 110.
Azay-le-Rideau, 13.

Bacillus salmonis-pestis, 205.
Barberry, 72.
Batsford, 132.
Bayley Balfour, Sir Issac, 114, 202, 224, 236.
Bean, W. J., 61, 63, 64, 69, 72, 97, 176, 188.
Beaumont, 31.
Bellingham, 31.
Benhadad, King of Syria, 11.
Bentham, 81, 142.
Bernina, 164.
Bird's-Foot Trefoil, 219.
Bitton Vicarage, 28, 168, 221, 235.
Boscawen, Rev. A. T., 88.
Botrytis, 23, 204, 206.
Bowles, E. A., 10, 154, 157.
Broom, 133.
Bugbane, 154, 155.
Bugwort, 36.
Bulbs, 1-26, 140.
Butcher's Broom, 134.

Caerhays, 114, 142.
Calico Bush, 65.
Campernelle, 7.
Cardross, 216.
Castle Kennedy, 109.
Celandine, 192.
Channel Islands, 75.
Chatham Islands, 83, 84, 86.
Cheeseman, 84, 87, 88, 89, 156.
Chenonceaux, 29.
Cherry-Laurel, 86, 135.
Chile, 71-73, 76-81.
Choate, Joseph, 59.
Christmas Rose, 155.
Colchicum, 10, 13.
" Contractile " Roots, 2.
Corms, 1, 2, 3.
Cornflowers, 50.
Cornwall, 56, 80, 88.
Correvon, Mons. H., 37.
Corswall, 80.
Crabbe, George, 220.
Craighlaw, 18.
Crocus, 1, 4, 8, 9, 11.
Crowfoot, 219.
Cypress, 48.

Daffodil, 7, 33.
Darwin, 42, 72, 160, 225.
Dawyck, 105.
Diane de Poictiers, 29.
Dogtooth Violet, 16.
Dogwood, 95.
Dorrien Smith, 4.
Douglas, David, 66, 70.

General Index

Doves' Dung, 11.
Dutch Crocuses, 9.

Edinburgh Botanic Garden, 63, 69, 94, 133, 161, 198, 200, 202, 224.
Eley, Charles, 45.
Ellacombe, Canon, 28, 132, 168, 191, 221, 235.
Elwes, H. J., 12, 23, 165, 225, 235.
Evelyn, John, 47.
Evening Primrose, 51.

Falstaff, 222.
Farrer, Reginald, 39, 66, 102, 152, 154, 182, 188, 196, 201, 203.
Fochabers, 70.
Forget-me-not, 136, 219, 220.
Forrest, 14, 66, 102, 113, 202.
Foxglove, 51, 136, 159, 160.
Fritillary, 13, 14.

Glasnevin Botanic Garden, 71, 92, 162.
Glasserton, 75.
Globe Artichoke, 162, 222.
Globe Flower, 158.
Glory-of-the-Snow, 15.
Goat's Beard, 215.
Golden Drop, 196.
Graham, Colonel, 31.
Grant, Allen, 9.
Gray, S., 219.
Ground Pine, 219.
Grove, A., 21, 24 n., 152, 206.
Gulf Stream, 53.

Haughton, Dr., 55.
Hawthorn, 58, 60.
Heath Family, 102, 112.
Henry, Dr., 102.
Hoare, A. H., 125.
Holly, 47.
Honeysuckle, 167.
Hooker, Sir Joseph, 56, 85, 92, 109, 142, 156, 228.
Hope, Miss Frances, 42, 233.
Horse-Chestnut, 58.

Hound's-Tongue, 213.
Houseleeks, 186.
Huxley, T., 205.
Hyacinth, 219, 225.

Iffley, 13.
Iris, 13, 143.

Jessamine, 167.
Jonquil, 7.
Judas Tree, 62.

Kapuka, 85.
Kew Gardens, 63, 69, 72, 83, 85, 97, 101, 106, 156, 169, 176, 188, 231.
King-cup, 159.
Kingdon Ward, 66.
Knotweed, 144.

Laburnum, 58, 61.
Lady's Slipper, 40-42.
Lady's-Smock, 150.
Larkspur, 44.
Laurustinus, 90.
Leny, 108.
Leonardslee, 71, 235.
Levens, Manor of, 30.
Lily, 1, 13, 17-26, 143, 227.
Lily-of-the-Valley, 2, 151.
Limax agrestis, 203.
Linnaeus, 225, 226.
Lobelia, 36.
Loch Fyne, 80.
Logan, 137.
Lowe, Dr., 233.
Lubbock, Sir John, 98.
Lungwort, 150.
Lyte, Henry, 220.

Mangles, Mr., 9.
Manzanita, 133.
Marsh Marigold, 159.
Mary, Queen of Scots, 30.
May, 59, 60.
Meadowsweet, 215.
Merodon Equestris, 8.
Mexican Orange-Flower, 69.

General Index

Michaelmas Daisy, 214.
Mignonette, 50.
Millais, J. G., 57.
Mochrum Castle, 120.
Monkshood, 138.
Monreith, 45, 46, 61, 78, 90, 92, 151, 183, 200.
Montenegro, 15, 141.
Mulberry at Hatfield, 30.
Mullein, 136.
Myrtle, 75.

Napier Shaw, Sir, 55 n.
Narcissus, 4, 5, 6, 7, 8, 11.
New Zealand Shrubs, 83-90.

Onion Tribe, 160.
Orchids, 37, 39.

Pansy, 190.
Passion Flower, 177.
Pearl Bush, 92.
Penzance, 86, 201 n.
Peter the Great, 48.
Pheasants, 74.
Pheasant's-Eye Narcissus, 7.
Plane, 219.
Pollok, 3.
Poolewe, 57.
Poppy, 50, 51, 81, 212.
Portugal Laurel, 135.
Potash in wood ash, 24.
Primrose, 136.
Primula, 136-138.
Prior, Dr., 158, 230.
Privet, 135.
Procopius, 55.
Purdy, Carl, 141.

Queen-of-the-Prairie, 215.

Rabbits, 5, 9, 13, 67, 128-130, 135, 138, 157, 194, 210, 212, 214.
Raphides, 8.
Rhizomes, 1, 2, 3.
Rhizopus necans, 19.
Rhododendrons, 57, 102-126, 137.

Robinson, W., 28, 42, 48, 128, 177, 217.
Rockcress, 217.
Rockfoil, 217.
Rock Rose, 185.
Rose, 167, 179.
Rosebay, 231.
Rose Campion, 218.
Ross, Sir John, 78, 86.
Ruskin, John, 166.

Sargent, Prof., 116.
Saxifrage, 190.
Sayes Court, 48.
Scilly Isles, 3, 4, 75, 88.
Scott, Sir Walter, 46.
Scorpion Grass, 220.
Sea-Holly, 22.
Shepherd's Weather Glass, 81.
Slugs, Destruction of, 157.
Smith, Prof. W. W., 94.
Snowdrop, 1, 3, 4, 5, 8, 13.
Snowflake, 5.
Spiny Madwort, 150.
Squill, 4, 11, 15, 16.
Stair, Earl of, 234.
Star of Bethlehem, 11, 12.
Stephanitis rhododendri, 124.
Stewart, L., 133 n., 200.
St. John's Wort, 185.
Stonecrop, 186, 191.
Sunflower, 164.
Sweet Peas, 232.

Temperature on West Coast, 53-55.
Tristram, Canon, 12.
Tubers, 1, 2, 3.
Tulip, 1, 4, 8, 16, 17.

United States, Importation of Plants into, 19.

Viola, 190.
Virginian Fringe Tree, 68.
Virgin's Bower, 217.

Watson, W., 119.
Wayfaring Tree, 91.

General Index

Weathers, Mr., 7 n., 195.
Webster, A. D., 40.
Whitebeam, Himalayan, 100.
Wild Garlic, 160.
Wilks, Rev. W., 51.
Willow Gentian, 136, 215.
Willow Herb, 201.
Wilson, E. P., 14, 66, 91, 97, 99, 102, 112, 116, 175, 177.
Wilson, Miss Mary, 56.

Wildflower, 217.
Winter Aconite, 2.
Wistaria, 58, 167.
Witch Hazel, 96.
Wyber, 216.

Yellow Bugle, 219.
Yellow Flowers, Preponderance of, 9.
Yew, 31, 32, 46, 47.

Index of Plant Names

Abelia floribunda, 168, 178.
Abutilon megapotamicum, 79, 168, 174.
    vitifolium, 78, 168, 174.
Acantholimum glumaceum, 204.
    venustum, 204.
Acer pseudo-platanus, 219.
Acaena, 214.
Aconitum, 138.
    Chinense, 138.
    Fischeri, 138.
    Japonicum, 138.
    napellus, 138.
    Wilsoni, 36, 138.
Adenophora Lamarckii, 149.
    megalantha, 149, 203.
Adonis Amurensis, 149.
    vernalis, 149.
Aethionema, Miss Wilmott, 190.
Ajuga chamaepitys, 219.
Allium, 221, 234.
    acuminatum, 161.
    Beesianum, 161.
    Karataviense, 161.
    narcissiflorum, 160.
    Pedemontanum, 160.
    sphaerocephalum, 161, 234.
    sub-hirsutum, 161.
    ursinum, 160.
Aloysia citriodora, 179.
Alstroemeria, 12.
    Hookeri, 165, 235.
Alyssum saxatile, 185.
    spinosum, 150, 151.
Amaryllis belladonna, 26, 176.
Ampelopsis, 166.

Anagallis arvensis, 81.
Anchusa myositidiflora, 150.
    Italica, 149.
Anemone, 139, 217.
    alpina, 139.
    Appenina, 36, 139.
    blanda, 139.
    hepatica, 139.
    Japonica, 35, 213, 214.
    narcissiflora, 152.
    palmata, 194, 195.
    rivularis, 152.
    rupicola, 152.
    sulphurea, 139.
Anemopsis, 152 n.
Anemonopsis macrophylla, 152.
Antholiza paniculata, 214.
Arctostephalos manzanita, 69, 133.
Arenaria Balearica, 156.
    montana, 185.
Aristolochia, 172.
Artemisia, 221.
Arum, 221.
Aster alpinus, 149.
    Fremonti, 149.
    Lipskii, 224.
    "St. Egwin," 36.
    salsuginosus, 149.
    sub-caeruleus, 149.
Astilbe Davidi, 139, 215.
Aubretia, 184, 185, 217.
    deltoidea, 33.
    "Dr. Mules," 185.
    Graeca, 33, 185.
    "Mrs. Lloyd Edwards," 185.

Index of Plant Names

Azalea, 110, 116.
Azara microphylla, 73.

Belladonna, 221.
Benthamia fragifera, 95.
Berberidopsis, 172.
    corallina, 35, 172.
Berberis, 110.
    aquifolia, 75, 148.
    buxifolia, 72, 74.
    Darwini, 72, 225.
    dulcis, 74.
    empetrifolia, 72.
    pinnata, 75, 148.
    stenophylla, 72.
    vulgaris, 73.
Boenninghousenia albiflora, 165.
Brodiaea volubilis, 196.
Buddleia, 82, 93.
    alternifolia, 94.
    Colvillei, 92.
    globosa, 34, 35, 82.
    variabilis, 93, 94.
Buphthalmum speciosum, 139.

Caesalpinia Gillesii, 169, 170.
    Japonica, 170.
Calandrinia grandiflora, 196.
    umbellata, 196.
Calceolaria violacea, 82, 107.
Callistemon speciosus, 88.
Calluna vulgaris, 132.
    vulgaris Alporti, 132.
Caltha, 159.
    palustris, 159.
Camassia Cusicki, 148.
    Leichtlini, 148.
Camelia reticulata, 174.
Campanula amabilis, 151.
    barbata, 158.
    lactiflora, 136, 212.
    latifolia, 136, 158.
    modesta, 187.
    muralis, 231.
    Portenschlagiana, 190, 231.
    pusilla, 187.
    rhomboidalis, 158.
    Sarmatica, 158.

Cardamine trifolia, 150.
Carmichaelia australis, 90.
Carophyllaceae, 234.
Carpenteria, 71.
    Californica, 71, 178.
Ceanothus rigidus, 178.
    thyrsiflorus, 178.
    Veitchianus, 178.
Centaurea macrocephala, 139, 164.
    Rhaponticum, 139, 163.
Cerastium tomentosum, 35.
Cercis siliquastrum, 62.
Cheimonanthus fragrans, 68.
Chelidonium Franchetianum, 192.
Chionanthus, 68.
    Virginica, 68.
Chionodoxa, 15, 20.
    gigantea, 15.
    Luciliae gigantea, 231.
Choysia ternata, 69.
Cimicifuga, 154.
    Americana, 155.
    cordifolia, 155.
    Dahurica, 155.
    foetida, 155.
    Japonica, 155.
    racemosa, 155.
    simplex, 36, 155.
Cistus, 62.
    Corbariensis, 63.
    crispus, 64.
    Cyprius, 63, 64.
    ladaniferus, 63, 64.
    laurifolius, 63, 64, 65.
    Loreti, 63.
    populifolius, 63.
    purpureus, 64.
    recognitus, 64.
    villosus, 64.
Clematis, 174, 176, 217.
    Armandi, 175.
    Fargesi, 175.
    flammula, 175.
    florida, 176.
    indivisa, 175.
    Jackmanii, 176.
    lanuginosa, 176.

Index of Plant Names

Clematis montana, 175.
    montana rubens, 175.
    montana Wilsoni, 175.
    patens, 176.
Clerodendron Fargesi, 95.
    foetidum, 95.
    trichotomum, 94, 95.
Clianthus puniceus, 89, 107, 169.
Codonopsis, 157.
    meleagris, 157.
    ovata, 157.
    sylvestris, 157.
    viridiflora, 157.
Colchicum, 10.
    speciosum, 11.
Compositae, 163, 198, 199.
Convallaria, 218.
Convolvulus, 172.
    althaeoides, 215.
    batatas, 222.
Cornus capitata, 95.
    Florida, 59, 95.
    kousa, 95.
    mas, 95.
Corokia cotoneaster, 148.
    virgata, 148.
Cotoneaster horizontalis, 199.
Crataegus, 65.
    Carrierei, 65.
Crinodendron, 76.
    Hookeri, 76.
Crinum Powelli album, 35.
Crocus longiflorus, 10.
    speciosus, 10, 35.
    vernus, 9.
    zonatus, 10.
Cryptomeria Japonica, 216.
Cupressus Lawsoniana, 48.
    macrocarpa, 48.
    Nootkatensis, 48.
Cyananthus lobatus, 196.
Cynara scolymus, 162.
Cynoglossum, 198.
    amabile, 213.
    Appeninum, 149, 198.
    nervosum, 35, 148, 198.

Cypripedium calceolus, 39, 40, 41, 43.
    spectabile, 42, 43.
Cytisus albus, 133.
    praecox, 133.

Desfontainea, 80.
    spinosa, 79, 166.
Deutzia longifolia, 99.
    scabra, 99.
Dianthus, 234.
    alpinus squarrosa, 186.
    callizonus, 186.
    deltoides, 186.
    neglectus, 186, 190.
Dierama pulcherrimum, 161.
Digitalis ambigua, 160.
Diplacus, 156.
Drosera, 122.
Dryas lanata, 185.
    octopetala, 185.

Eccremocarpus scaber, 172.
Echinacea purpurea, 199.
Embothrium coccineum, 82, 204.
Epilobium obcordatum, 201.
    Dodonaei, 201.
    Fleischeri, 201.
Erica arborea, 73, 131.
    arborea alpina, 131.
    australis, 131.
    carnea, 132, 191.
    Darleyensis, 132.
    Mediterranea, 131, 132.
    stricta, 132.
    vagans, 132.
Erigeron mucronatus, 156, 211
    salsuginosus, 149.
Erinus, 187.
    alpinus, 187.
    alpinus carmineus, 187.
Eritrichium nanum, 196.
Eryngium maritimum, 222.
Eruca sativa, 223.
Erythronium, 141.
    Americanum, 16.
    californicum, 142.
    dens-canis, 16, 141.

Index of Plant Names

Erythronium revolutum, 16, 142.
   revolutum Johnsoni, 142.
Escallonia, 82.
   Exoniensis, 82.
   Langleyensis, 82, 148.
   macrantha, 82.
   Philippiana, 82.
   pterocladon, 82.
   punctata, 82.
   rubra, 82, 148.
Eschscholtzia Californica, 50.
Eucryphia, 77, 78, 120.
   Billarderi, 78.
   cordifolia, 78.
   pinnatifolia, 77, 78.
Eurybia Gunniana, 84.
Exochorda, 110.
   Giraldi, 92.
   grandiflora, 92, 107.

Fatsia Japonica, 25, 100.
Forsythia, 96.
   intermedia, 96.
   suspensa, 96.
Fremontia, 71, 173.
   Californica, 70, 173.
Fritillaria, 20.
   aurea, 14.
   imperialis, 14.
   meleagris, 13, 140.
   pallida, 14.
   pudica, 14.
   Pyrenaica, 14, 140.
Fuchsia, 228.
   corymbiflora, 173.
   fulgens, 173.
   globosa, 173.
   Riccartoni, 173.
Funkia Sieboldi, 138.

Galanthus, 20.
   nivalis, 3.
Gaultheria shallon, 132.
   Veitchiana, 132.
Garrya, 70.
   elliptica, 70.
Gaya Lyalli, 89.
Genista virgata, 101.

Gentiana acaulis, 190, 197.
   asclepiadea, 34, 136, 215.
   cruciata, 149.
   Farreri, 149, 197.
   Lagodeshiana, 149.
   macrophylla, 149.
   ornata, 149, 197.
   phlogifolia, 149.
   Przewalskyi, 229.
   septemfida, 149.
   Sino-ornata, 149.
   verna, 196.
Geranium Armenum, 139.
   Ibericum, 139.
   pratense, 36.
   sanguineum, 81.
   Wallichianum, 35.
Geum Borisii, 35.
Godetia, 50.
Griselinia littoralis, 85, 86.

Habenaria bifolia, 37, 42.
Habranthus pratense, 34, 82.
Hamamelis arborea, 96, 148.
   Japonica, 96, 148.
   mollis, 96, 148.
Helianthemum, 185.
Helianthus, 215.
   mollis, 164.
   multiflorus, 215.
Helicodiceros crinitus, 52.
Helleborus, 155.
Hippeastrum pratense, 34, 35, 82.
Houttuynia, 152 n.
Hyacinthus, 221.
   amethystinus, 225.
Hypericum fragile, 185.
   Hookerianum, 94.
   oblongifolium, 94 n.
   patulum, 94.
   patulum Henryi, 94.
   repens, 149, 185.
   reptans, 149, 185.

Indigofera, 99, 173.
   Gerardiana, 99, 173.
Inula grandiflora, 139.
   Roylei, 139.

Index of Plant Names

Iris chrysographes, 35.
   reticulata, 13.
   xiphium, 13, 143.
   xiphoides, 13.

Kalmia latifolia, 65, 133.
Kirengeshoma palmata, 36.
Kniphofia, 227.
   aloides, 35.

Laburnum alpinum, 62.
   alpinum Watereri, 62.
   vulgare, 62.
Lapageria, 172, 174.
   rosea, 172.
Lardizabala biternata, 171.
Lathyrus, 170.
   Drummondi, 171.
   grandiflorus, 171.
   latifolius, 171.
   pubescens, 170.
   rotundifolius, 171.
Lavatera Olbia, 139.
Lepachys pinnata, 36.
Leptospermum ericoides, 90.
   lanigerum, 90.
   scoparium, 90.
   scoparium Chapmanii, 90.
   scoparium Nicholli, 90.
Leucojum vernum, 5.
   vernum Vagneri, 5.
Libertia, 233.
   formosa, 139.
   grandiflora, 35, 139.
Lilium auratum, 18, 24.
   Browni, 25, 205, 228.
   candidum, 24.
   Chalcedonicum, 23.
   croceum, 22, 36.
   Davuricum, 22.
   Davuricum incomparabile, 22, 144.
   Davuricum luteum, 22.
   giganteum, 1 n., 21, 23, 25, 143.
   Henryi, 25.
   Humboldti, 206.
   martagon, 22, 143.
   martagon Dalmaticum, 22.
   monadelphum, 21, 23.
   monadelphum szovitzianum, 22.
   pardalinum, 23, 25, 143.
   Parryi, 207.
   Philadelphicum, 207.
   pomponium, 24.
   pyrenaicum, 23, 25, 143.
   regale, 22, 25.
   rubellum, 207.
   Sargentiae, 22, 107.
   speciosum, 24.
   superbum, 23, 25.
   testaceum, 23, 36.
   tigrinum, 23, 26.
   tigrinum Fortunei, 23.
   tigrinum splendens, 23.
   umbellatum, 22.
   umbellatum incomparabile, 22.
Lindelofia spectabilis, 148.
Linnaea borealis, 226.
Linum capitatum, 196.
Lippia citriodora, 179.
Lithospermum, 188.
   graminifolium, 185.
   prostratum, 185.
   prostratum " Heavenly Blue," 185.
Lobelia, 36, 82.
   cardinalis, 82.
   fulgens, 82.
   splendens, 82.
   tupa, 82.
Lonicera, 221.
   fragrantissima, 178.
   Standishi, 178.
   tragophylla, 177.
Lotus corniculatus, 219.
Lychnis, 234.
   diurna, 218.
Lysimachia nummularia, 85.

Magnolia, 96.
   conspicua, 96.
   grandiflora, 179.
   parviflora, 96.
Mahonia fascicularis, 75.
Malvastrum coccineum, 196.
Mandragora officinarum, 234.

Index of Plant Names

Mazus, 156.
Meconopsis Cambrica, 81 n., 212.
   integrifolia, 52.
   latifolia, 52, 186.
   Nepalensis, 52.
   paniculata, 52.
   Pratti, 52.
   punicea, 52.
   simplicifolia, 51, 186, 197, 212 n.
   Wallichiana, 52.
Mertensia echioides, 198.
   elongata, 198.
   lanceolata, 198.
   Siberica, 198, 226.
   Virginica, 198.
Metrosideros, 88.
   diffusa, 88.
   lucida, 88.
   robusta, 88.
   tomentosa, 88.
Mimulus Bartonianus, 149, 155.
   cardinalis, 155, 156.
   glutinosus, 156.
   Lewisi, 149, 155, 156.
   radicans, 156.
Moltkia petraea, 188.
Montbretia, 214.
Muscari, 20.
   conicum, 36.
Mutisia, 199, 200, 201.
   clematis, 201 n.
   decurrens, 178, 199, 200, 201.
   ilicifolia, 201.
Myosotis, 220.
   rupicola, 196.
   sylvatica, 136.
Myrtus communis, 75.
   luma, 76, 121.

Narcissus, 6, 8.
   Barri conspicuus, 6.
   bicolor, 6.
   bulbocodium, 7.
   bulbocodium citrinus, 7, 36.
   cyclamineus, 7, 140.
   Johnstoni, 6, 33.
   jonquilla, 7.
   minimus, 6.
   minor, 6, 15.
   nanus, 15.
   odorus, 7.
   poeticus, 7, 140.
   poeticus ornatus, 6 n., 7.
   pseudo-narcissus, 6, 33, 34.
   tazetta, 6, 7 n.
   triandrus, 6, 7, 34.
   " Will Scarlett," 6 n.
Nerine Bowdeni, 26.
Notospartium Carmichaeliae, 90.

Oenothera, 51.
Olearia, 83, 85.
   Buchanani, 83.
   Chathamica, 84.
   haasti, 83.
   ilicifolia, 83.
   insignis, 84.
   macrodonta, 83, 148.
   nitida, 83, 107.
   nummularifolia, 85.
   semidentata, 84.
   stellulata, 84, 107.
   Traversi, 83, 84, 148.
Omphalodes Cappadocica, 150.
Onosma albo-roseum, 196.
   Tauricum, 196.
Ophrys, 37.
Orchis, 37.
   foliosa, 42.
   latifolia, 37, 42.
   maculata, 37, 42.
   maculata superba, 42.
   mascula, 42, 81, 160.
   pyramidalis, 37, 42.
Oreodaphne Californica, 227.
Ornithogalum Arabicum, 11.
   nutans, 11, 140.
   pyramidale, 11.
   thyrsoides, 11.
   umbellatum, 11.
Osmanthus Delavayi, 99, 148, 206.
   ilicifolius, 148.
Oxalis, 213.

Index of Plant Names

Paeonia Cambessodessi, 207.
  lutea, 149.
  Mlokosewitchii, 149, 229.
Papaver aculeatum, 149.
  nudicaulis, 212.
  umbrosum, 36, 149.
Passiflora coerulea, 177.
  " Constance Elliot," 177.
Penstemon confertum, 196.
  cordifolius, 168.
  heterophyllus, 196.
Phacelia campanularia, 50.
Philadelphus, 36, 69, 110.
  microphyllus, 69.
Phlox subulata, 184.
  verna, 185.
Phormium tenax, 45.
Phygelius Capensis, 101.
Phyteuma comosum, 203.
Pieris floribunda, 99, 148.
  formosa, 121.
  Japonica, 99, 148.
  Tawaniensis, 99.
Piperaceae, 152 n.
Piptanthus Nepalensis, 99, 107.
  tomentosus, 99.
Plagianthus Lyalli, 89.
Platanus orientalis, 219.
Podophyllum Emodi, 148.
  peltatum, 148.
Polygonaceae, 209.
Polygonum, 144, 145.
  affine, 35.
  alpinum, 145, 210.
  Auberti, 210.
  Baldschuanicum, 144, 210.
  bistorta, 210.
  Brunonis, 35.
  campanulatum, 145, 210.
  cuspidatum, 210.
  equisetiforme, 145, 210.
  polystachyum, 144, 145, 209, 210.
  Sachalinense, 144, 145, 209.
  sphaerostachyum, 210.
  vaccinifolium, 144, 211.
Potentilla " Gibson's Scarlet," 34.
  Tonguei, 201.

Primula, 152.
  Beesiana, 153.
  Bulleyana, 137, 153, 154.
  Burmannica, 137, 153.
  Cashmeriana, 153.
  Cockburniana, 154.
  denticulata, 153.
  helodoxa, 137, 138, 153.
  Japonica, 44, 137, 153, 154.
  Poissoni, 153.
  pulverulenta, 137, 138, 153, 154.
Prostanthera rotundifolia, 107.
Prunella grandiflora, 192.
  Webbiana, 192.
Prunus lauro-cerasus, 86.
Pulmonaria angustifolia, 149, 150.
  angustifolia rosea, 150.
  Arvernensis, 149.
  saccharata, 150.
Pyrus vestita, 100.

Ramondia Pyrenaica, 186.
Ranunculaceae, 175, 209.
Ranunculus acris, 219.
  repens, 219, 225.
Rhododendron, 102, 221.
  adenogynum, 103.
  aeruginosum, 124.
  ambiguum, 107.
  arboreum, 57, 106, 107, 109, 117, 121, 137, 202.
  argyrophyllum, 103.
  " Ascot Brilliant," 117.
  Aucklandi, 93, 103, 107, 118, 228.
  Augustini, 103, 107, 130.
  auriculatum, 124.
  barbatum, 105, 106, 107, 117, 121, 122, 123.
  " Beauty of Bagshot," 117.
  Broughtoni, 117.
  bullatum, 103.
  calophytum, 103, 107.
  campanulatum, 57, 106, 107, 108.
  Catawbiense, 116.
  Caucasicum, 116, 121.
  ciliatum, 36, 106, 121.
  cinnabarinum, 121, 155.

Index of Plant Names

Rhododendron cinnabarinum Roylei, 117.
  crassum, 103, 107.
  Davidi, 107.
  decorum, 103, 107, 124.
  discolor, 124.
  Edgeworthi, 107, 174.
  erastum, 102.
  Falconeri, 121, 124, 228.
  Fargesi, 103, 106, 107.
  ferrugineum, 121.
  Fortunei, 107, 235.
  Fortunei cyanocarpum, 202.
  " George Hardy," 117.
  giganteum, 102.
  glaucum, 121, 130.
  Griffithianum, 93, 103, 123, 228, 235.
  habrotichum, 123.
  haematochilon, 106, 107.
  haematodes, 103, 124.
  hippophaeoides, 124.
  hirsutum, 112, 114, 121.
  Hodgsoni, 121, 228.
  Indicum, 121.
  Kamschaticum, 202.
  " Kate Waterer," 117.
  Kewense, 107, 117.
  " Lady Clementine Mitford," 117.
  " Lady Eleanor Cathcart," 117.
  Loderi, 117, 235.
  " Loder's White," 117.
  lutescens, 107.
  Maddeni, 117.
  maximum, 124.
  megacalyx, 174.
  micranthum, 124.
  " Minnie," 117.
  Moupinense, 148.
  mucronulatum, 148.
  neriiflorum, 103.
  Nobleanum, 124.
  orbiculare, 103.
  oreodoxa, 103, 104, 106.
  oreotrephes, 103.
  pachytrichum, 107, 121.
  parvifolium, 148.
  pholidotum, 124.
  " Pink Pearl," 117.
  ponticum, 116, 118, 119, 120, 121, 135.
  praecox, 130, 148.
  " Sappho," 117.
  Schlippenbachii, 107.
  Scottianum, 107, 174.
  sinogrande, 103, 107, 121, 124.
  Smirnovi, 121.
  Soulei, 103, 107, 120.
  strigillosum, 107.
  Sutchuenense, 107.
  Thomsoni, 106, 107, 117, 121, 228.
  Thomsoni eximium, 124.
  viscosum, 124.
  Yunnanense, 103.
Rhus, 221.
Ribes Americanum, 67.
  aureum, 67.
  Gordonianus, 67.
  Missouriensis, 67.
  sanguineum, 66, 67.
  speciosus, 178.
Robinia hispida, 170.
  neo-Mexicana, 107.
Romneya Coulteri, 44.
Rosa Moyesii, 98.
Rosa " Climbing Papa Gontier," 179.
Roscoea, 73, 163.
  " August Beauty," 163.
  capitata, 163.
  cautlioides, 163.
  Humeana, 163.
  purpurea, 163.
Rubus deliciosus, 70.
  Nutkanus, 70.
  spectabilis, 70.
Rudbeckia maxima, 164.
  pinnata, 36.
  laciniata, 164.
Ruscus aculeatus, 134.

Salvia, 73.
  nemorosa virgata, 36.
Saxifrage aizoon rosea, 44, 190.
  Andrewsii, 190.
  apiculata, 36, 149, 190.

Index of Plant Names

Saxifrage Bathoniensis, 149.
   Beesiana, 107, 149, 150.
   Burseriana, 149.
   cotyledon, 186.
   cotyledon Icelandica, 224.
   crassifolia, 149.
   cymbalaria, 187.
   Delavayi, 107, 149, 150.
   " Guildford Seedling," 149.
   Irvingi, 186.
   juniperina, 149.
   longifolia, 186.
   Megasea group of, 107, 149.
   muscoides, 149.
   " Red Admiral," 149.
   sancta, 149.
   Sibthorpi, 187.
   Stracheyi, 107, 150.
Scilla, 15, 20.
   bifolia, 15, 190.
   campanulata, 16, 141.
   Hispanica, 16, 141, 225.
   nutans, 140.
   Siberica, 15, 190.
   verna, 16.
Sedum, 186.
   acre, 191.
   Anglicum, 191.
   elegans, 192.
   Guatamalense, 191.
   rupestre, 191.
   spurium, 191.
Selaginella stolonifera, 217.
Senecio, 86, 221.
   compactus, 86, 148.
   Greyi, 86, 148.
   Hectori, 86, 87.
   Hunti, 86, 148.
   Jacobaea, 86.
   Kirkii, 86.
   Munroi, 148.
Silene alpestris, 186.
   Elizabethae, 196.
   laciniata Purpusi, 196.
   Schafta, 192.
Solanum crispum, 171.
   jasminoides, 171.

Solanum tuberosum, 222.
Sophora tetraptera, 87.
   tetraptera microphylla, 87.
Spiraea, 98, 110.
   Aitchesoni, 97.
   arborea, 97.
   Arguta, 96, 107.
   aruncus, 139, 215.
   brachybotris, 97.
   bracteata, 96, 97, 148.
   canescens, 97.
   decumbens, 185.
   Douglasi, 97.
   Henryi, 97.
   Japonica, 98.
   Lindleyana, 97.
   lobata, 139, 215.
   palmata, 139.
   Van Houttei, 96, 148.
Stenanthium robustum, 202.
Stephanotis, 172,
Sternbergia, 195.
Styrax Hemsleyanum, 100.
   Japonicum, 100.
   Obassia, 100.

Tecoma radicans, 168.
Thalictrum dipterocarpum, 27.
Thuya plicata, 48.
Thymus serpyllum coccineum, 190.
   serpyllum micans, 190.
Tricuspidaria, 76, 77, 120, 168.
   dependens, 77, 148.
   lanceolata, 76, 148, 167.
Trillium grandiflorum, 43, 138.
Tritonia, 214.
   Pottsi, 214.
Trollius, 159.
   Asiaticus, 159.
   Europaeus, 159.
   " Orange Crest," 159.
   patulus, 158.
   " Y. Smith," 159.
   Yunnanensis, 159.
Tropaeolum speciosum, 71, 171, 178.
Tulipa Clusiana, 16.
   Greigi, 16.

Index of Plant Names

Tulipa Kaufmanniana, 17.
   Persica, 17.
   praestans, 16, 36.
   sylvestris, 17, 142.
   Tubergiana, 16.

Umbellularia Californica, 227.

Vaccinium corymbosum, 138.
Verbascum, 136.
Veronica ambigua, 148, 213.
   cupressoides, 148.
   filifolia, 157.
   filiformis, 157.
   longifolia sub-sessilis, 148.
   parviflora, 87.
   salicornis, 148.
   salicifolia, 87.
   spicata, 148.
   Traversi, 87.
   Virginica, 148.

Viburnum, 110.
   Carlesii, 90, 91 *n*., 148.
   crassifolium, 92, 148.
   fragrans, 91 *n*., 148
   lantana, 91.
   rhytidophyllum, 91, 148.
   tinus, 90.
   tinus lucidum, 90.
   tomentosum, 91.
   tomentosum Mariesii, 91, 148.
   tomentosum plicatum, 91.
   utile, 92, 148.
Vittadenia triloba, 211.
Vitis inconstans, 166, 174.

Wistaria, 167, 172.
   Chinensis, 170.
   multijuga, 170.

Xanthoceras sorbifolia, 92.

www.ingramcontent.com/pod-product-compliance
Lightning Source LLC
Chambersburg PA
CBHW061935290426
44113CB00025B/2917